This is a circle.
	 This is a square.
circle	 This is a circle. / square.
yes	 Is this a square? yes / no
yes	 This circle has 3 parts. Are all 3 parts the same size? yes / no
3	Therefore we say that the circle has ___ equal parts.
4	 This square has ___ equal parts.
5	 This stick has ___ equal parts.

FINAL EXAM - Side 2, contd.

$9 \div \frac{3}{5} = ?$ $\qquad\qquad$ $\frac{4}{9} \div 8 = ?$

$$\begin{array}{r} 13 \\ -\ 5\frac{5}{7} \\ \hline \end{array} \qquad\qquad \begin{array}{r} 173 \\ -\ 2\frac{4}{13} \\ \hline \end{array}$$

Change to improper fractions:

$2\frac{5}{9} =$ $\qquad\qquad$ $4\frac{3}{5} =$

$1\frac{1}{3} \times 2\frac{1}{2} = ?$ $\qquad\qquad$ $4\frac{1}{8} \div 2\frac{1}{4} = ?$

$\frac{2}{5} + \frac{4}{7} = ?$ $\qquad\qquad$ $\frac{5}{6} - \frac{2}{5} = ?$

$1\frac{1}{2} + \frac{2}{3} = ?$ $\qquad\qquad$ $2\frac{1}{4} - \frac{7}{8} = ?$

← This is one of the three equal parts of this circle.

"one of three" → $\frac{1}{3}$

---

four

← This is one of the three / four equal parts of this square.

$\frac{1}{4}$

"one of four" → $\frac{\quad}{4}$

This is one of the five equal parts of this stick.

$\frac{1}{5}$

"one of five" → $\frac{1}{\quad}$

$\frac{1}{6}$

"one of six" → $\frac{1}{\quad}$

$\frac{1}{8}$

"one of eight" → $\frac{\quad}{8}$

$\frac{1}{9}$

"one of nine" → ——

Answers	Frame
$\frac{8}{48}$	$\frac{1}{6} + \frac{5}{8} = ?$ The lowest common denominator is 24, but let's use 48 (8 × 6).
$\frac{30}{48}$	$\frac{1}{6} = \frac{\quad}{48}$ $\frac{5}{8} = \frac{\quad}{48}$
$\frac{38}{48} = \frac{19}{24}$	$\frac{8}{48} + \frac{30}{48} = \underline{\quad} = \underline{\quad}$
$\frac{3}{4} = \frac{30}{40}$ $\frac{3}{10} = \frac{12}{40}$ $\frac{30}{40} - \frac{12}{40} = \frac{18}{40}$ $\frac{18}{40} = \frac{9}{20}$	$\frac{3}{4} - \frac{3}{10} = ?$ Use 40 as the common denominator (instead of the lowest common denominator 20).

FINAL EXAM - Side 2

$$5\tfrac{6}{19} + 3\tfrac{7}{19} = \underline{\qquad}$$

$$25\tfrac{8}{15} + 6\tfrac{3}{15} = \underline{\qquad}$$

$$21\tfrac{6}{7} - 17\tfrac{2}{7} = \underline{\qquad}$$

Change to mixed numbers:

$\frac{7}{4} =$ $\qquad$ $\frac{13}{3} =$ $\qquad$ $\frac{19}{4} =$

$8 \times \frac{7}{12} =$ $\qquad$ $\frac{3}{7} \times 28 =$

Here are two of the three equal parts of this circle.

"two of three" ⟶ $\frac{2}{3}$

$\frac{2}{5}$

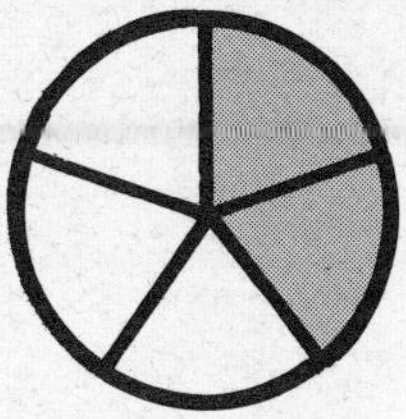

"two of five" ⟶ $\frac{2}{\ }$

$\frac{3}{4}$

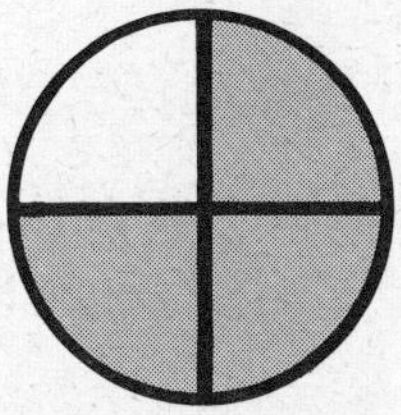

"three of four" ⟶ $\frac{\ }{4}$

$\frac{4}{7}$

"four of seven" ⟶ $\frac{4}{\ }$

$\frac{2}{9}$

"two of nine" ⟶ —

$\frac{5}{6}$

"five of six" ⟶ —

$\frac{4}{3}$	$1\frac{1}{3} - \frac{5}{12} = \ ?$
$\frac{16}{12}$	$1\frac{1}{3} = \frac{\quad}{3}$ $\frac{4}{3} = \frac{\quad}{12}$
$\frac{11}{12}$	$\frac{16}{12} - \frac{5}{12} =$
$2\frac{1}{2} = \frac{5}{2}$ $\frac{5}{2} = \frac{25}{10}$ $\frac{25}{10} - \frac{8}{10} = \frac{17}{10}$ $\frac{17}{10} = 1\frac{7}{10}$	$2\frac{1}{2} - \frac{4}{5} = \ ?$
$\frac{6}{24}$ $\frac{4}{24}$ $\frac{5}{12}$	We can always find a common denominator by multiplying the denominators. If this isn't the <u>lowest</u> common denominator, we will need to reduce our answer to lowest terms. $\frac{1}{4} + \frac{1}{6} = \ ?$ The lowest common denominator is 12, but let's suppose we multiplied 4×6 and used 24 as our common denominator: $\frac{1}{4} = \frac{\quad}{24}$ $\frac{1}{6} = \frac{\quad}{24}$ $\frac{6}{24} + \frac{4}{24} = \frac{10}{24} = \frac{\quad}{12}$

	These are <u>whole numbers</u>: 1 2 5 9 14 376
	These are <u>fractions</u>: $\frac{1}{2}$ $\frac{3}{4}$ $\frac{7}{9}$ $\frac{16}{25}$ $\frac{584}{957}$
2 5 14	Circle the whole numbers: 1 $\frac{1}{2}$ 2 $\frac{3}{4}$ 5 $\frac{7}{9}$ 14 $\frac{16}{25}$
$\frac{9}{10}$ $\frac{15}{37}$	Circle the fractions: $\frac{2}{5}$ $\frac{9}{10}$ 7 136 $\frac{15}{37}$
$\frac{5}{9}$ $\frac{4}{11}$ $\frac{27}{44}$	$\frac{5}{8}$ ← The top part of a fraction is called the <u>numerator</u>. Circle the numerators: $\frac{3}{7}$ $\frac{5}{9}$ $\frac{4}{11}$ $\frac{27}{44}$
$\frac{2}{3}$ $\frac{4}{5}$ $\frac{1}{20}$	The bottom part of a fraction is called the <u>denominator</u>. Circle the denominators: $\frac{7}{8}$ $\frac{2}{3}$ $\frac{4}{5}$ $\frac{1}{20}$

Answer	Problem
$\frac{11}{18} - \frac{4}{18} = \frac{7}{18}$	$\frac{11}{18} - \frac{2}{9} = ?$
$\frac{12}{15} - \frac{10}{15} = \frac{2}{15}$	$\frac{4}{5} - \frac{2}{3} = ?$
$\frac{9}{6}$	$1\frac{1}{2} + \frac{5}{6} = ?$ $1\frac{1}{2} = \frac{3}{2}$
$\frac{14}{6}$	$\frac{3}{2} = \frac{}{6}$ $\frac{9}{6} + \frac{5}{6} =$
$\frac{7}{3}$	Reduce to lowest terms: $\frac{14}{6} = \frac{}{3}$
$2\frac{1}{3}$	$\frac{7}{3} = \_\_\frac{1}{3}$
$\frac{22}{8} + \frac{5}{8} = \frac{27}{8}$ $\frac{27}{8} = 3\frac{3}{8}$	$2\frac{3}{4} + \frac{5}{8} = ? \quad \left(\frac{11}{4} + \frac{5}{8} = ?\right)$

fraction	A fraction / whole number always has a numerator and a denominator.
$\frac{1}{8}$ $\frac{3}{4}$ $\frac{11}{17}$	Circle the fractions: 8 14 $\frac{1}{8}$ 29 $\frac{3}{4}$ $\frac{11}{17}$
36 4	Circle the whole numbers: $\frac{59}{84}$ $\frac{121}{817}$ 36 4 $\frac{6}{7}$
top	The top / bottom part of a fraction is called the numerator.
$\frac{4}{9}$ $\frac{81}{109}$ $\frac{1}{2}$ (numerators circled)	Circle the numerators: $\frac{4}{9}$ $\frac{81}{109}$ $\frac{1}{2}$
$\frac{5}{6}$ $\frac{21}{100}$ $\frac{4}{19}$ (denominators circled)	Circle the denominators: $\frac{5}{6}$ $\frac{21}{100}$ $\frac{4}{19}$

$\frac{9}{21}$	$\frac{3}{7} + \frac{1}{3} = ?$
$\frac{7}{21}$	$\frac{3}{7} = \_\_$
$\frac{9}{21} + \frac{7}{21} = \frac{16}{21}$	$\frac{1}{3} = \_\_$
	$\frac{9}{21} + \_\_ =$
$\frac{1}{8} + \frac{6}{8} = \frac{7}{8}$	$\frac{1}{8} + \frac{3}{4} = ?$
40	$\frac{5}{8} - \frac{3}{5} = ?$
	What is the lowest common denominator? _____
$\frac{25}{40}$	$\frac{5}{8} = \_\_$
$\frac{24}{40}$	$\frac{3}{5} = \_\_$
$\frac{1}{40}$	$\frac{25}{40} - \frac{24}{40} =$
$\frac{4}{12}$	$\frac{1}{3} - \frac{1}{4} = ?$
$\frac{3}{12}$	$\frac{1}{3} = \_\_$
	$\frac{1}{4} = \_\_$
$\frac{4}{12} - \frac{3}{12} = \frac{1}{12}$	$\frac{4}{12} - \_\_ =$

TEST

fraction	$\frac{4}{7}$ is a fraction. / whole number.
yes	Is 316 a whole number? yes / no
23 150	Circle the whole numbers: 23 $\frac{4}{17}$ 150 $\frac{1}{8}$
$\frac{5}{8}$ $\frac{21}{39}$	Circle the fractions: $\frac{5}{8}$ 30 837 $\frac{21}{39}$
denominator	The bottom part of a fraction is called the numerator. / denominator.
$\frac{87}{316}$ $\frac{2}{3}$	Circle the denominators: $\frac{87}{316}$ $\frac{2}{3}$
$\frac{7}{12}$	Circle the numerator: $\frac{7}{12}$

Answer	Frame
	$\frac{1}{2} + \frac{1}{3} = ?$
6	What is the lowest common denominator? ____
$\frac{3}{6}$	$\frac{1}{2} = \frac{\ }{6}$
$\frac{2}{6}$	$\frac{1}{3} =$
$\frac{5}{6}$	$\frac{3}{6} + \frac{2}{6} =$
20	$\frac{1}{4} + \frac{3}{5} = ?$ What is the lowest common denominator? ____
$\frac{5}{20}$	$\frac{1}{4} =$ ___
$\frac{12}{20}$	$\frac{3}{5} =$ ___
$\frac{17}{20}$	$\frac{5}{20} + \frac{12}{20} =$
$\frac{10}{15}$	$\frac{2}{3} + \frac{1}{5} = ?$
	$\frac{2}{3} =$ ___
$\frac{3}{15}$	$\frac{1}{5} =$ ___
$\frac{13}{15}$	$\frac{10}{15} + \frac{3}{15} =$
$\frac{9}{18}$	$\frac{1}{2} + \frac{4}{9} = ?$
$\frac{8}{18}$	$\frac{1}{2} =$ ___
	$\frac{4}{9} =$ ___
$\frac{17}{18}$	$\frac{9}{18} + \frac{8}{18} =$

$\frac{1}{4}$

The denominator of the fraction tells us how many parts this circle has.

4	This circle has (2 / 4) parts.
	The numerator tells us how many of those parts we have shaded.
1	We have shaded (1 / 3) of the 4 parts.

$\frac{2}{5}$

denominator	The (numerator / denominator) tells us how many parts this circle has.
yes	Does the numerator tell us how many parts we have shaded? (yes / no)
5	We have shaded 2 of the __ parts of this circle.

	TEST
12	Find the lowest common denominator for each of the following pairs of fractions: $\frac{1}{3}$ and $\frac{1}{4}$ ____
14	$\frac{1}{2}$ and $\frac{2}{7}$ ____
24	$\frac{2}{3}$ and $\frac{1}{8}$ ____
9	$\frac{1}{9}$ and $\frac{1}{3}$ ____
15	$\frac{2}{5}$ and $\frac{4}{15}$ ____
$\frac{2}{8}$	$\frac{1}{4} = \frac{\quad}{8}$
$\frac{6}{9}$	$\frac{2}{3} = \frac{\quad}{9}$
15	$\frac{4}{15} + \frac{1}{5} = ?$ What is the lowest common denominator? ____
$\frac{3}{15}$	$\frac{1}{5} = \frac{\quad}{15}$
$\frac{7}{15}$	$\frac{4}{15} + \frac{3}{15} =$

3	 $\frac{2}{3}$ The denominator is 2. / 3.
2	The numerator is __.
3	 $\frac{3}{4}$ The numerator is 3. / 4.
4	The denominator is __.
$\frac{1}{2}$	Fill in the missing part of each fraction:
$\frac{1}{6}$	 $\frac{1}{\_}$  $\frac{\_}{6}$
$\frac{1}{3}$	
$\frac{2}{5}$	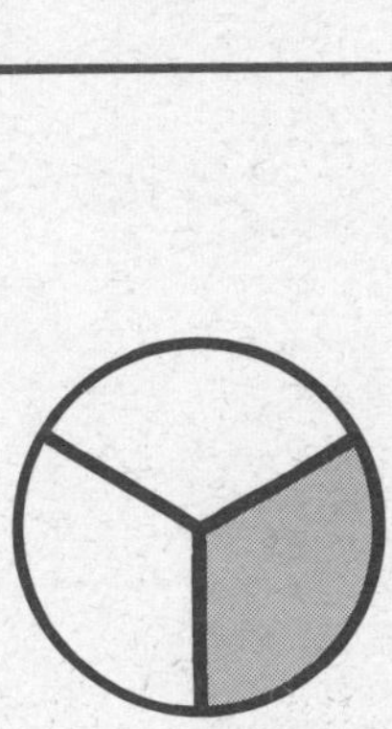 $\frac{\_}{3}$ 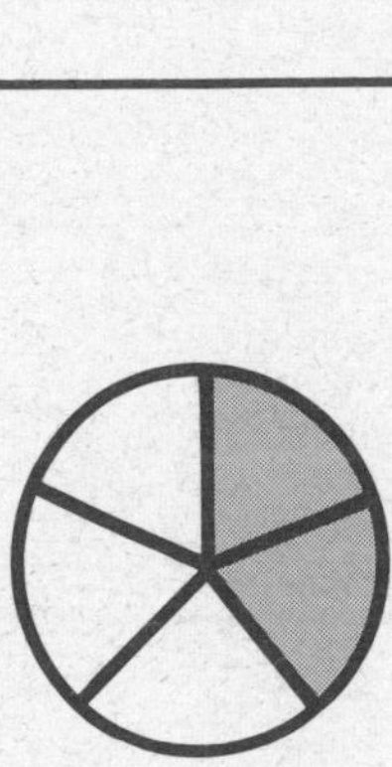 $\frac{2}{\_}$

Answers	Frames
$\frac{2}{6}$ $\frac{8}{20}$	$\frac{1}{3} = \frac{\quad}{6}$ (2 × 1) $\frac{2}{5} = \frac{\quad}{20}$ (4 × 2)
$\frac{9}{21}$ $\frac{30}{50}$	$\frac{3}{7} = \frac{\quad}{21}$ $\frac{6}{10} = \frac{\quad}{50}$
6 $\frac{4}{6}$ $\frac{5}{6}$	$\frac{2}{3} + \frac{1}{6} = ?$ What is the lowest common denominator? ____ Change $\frac{2}{3}$ to higher terms: $\frac{2}{3} = \frac{\quad}{6}$ So $\frac{4}{6} + \frac{1}{6} = \frac{\quad}{6}$
12 $\frac{4}{12}$ $\frac{5}{12}$	$\frac{1}{3} + \frac{1}{12} = ?$ What is the lowest common denominator? ____ $\frac{1}{3} = \frac{\quad}{12}$ So $\frac{4}{12} + \frac{1}{12} =$ ____
16 $\frac{2}{16}$ $\frac{3}{16}$	$\frac{1}{16} + \frac{1}{8} = ?$ What is the lowest common denominator? ____ $\frac{1}{8} = \frac{\quad}{16}$ $\frac{1}{16} + \frac{2}{16} =$

$\frac{2}{7}$

$\frac{2}{3}$

$\frac{3}{4}$

$\frac{3}{5}$

$\frac{3}{7}$

$\frac{4}{5}$

$\frac{4}{7}$

$\frac{5}{9}$

$\frac{2}{}$

$\frac{}{3}$

$\frac{3}{}$

$\frac{}{5}$

$\frac{3}{}$

$\frac{4}{}$

$\frac{}{7}$

$\frac{5}{}$

Answers	Frames
	Find the lowest common denominator (the lowest number divisible by both denominators) for each of the following pairs: $\frac{1}{6}$ and $\frac{1}{3}$ ___6___
20	$\frac{1}{20}$ and $\frac{1}{5}$ ______
12	$\frac{1}{4}$ and $\frac{1}{12}$ ______
14	$\frac{1}{7}$ and $\frac{1}{2}$ ______
15	$\frac{1}{3}$ and $\frac{1}{5}$ ______
divide	When we reduce a fraction, we (divide / multiply) both terms by the same number. To raise a fraction to higher terms, we multiply both terms by the same number. $\frac{1}{2} = \frac{?}{8}$
4	We multiplied the denominator 2 by ____ to get the new denominator 8.
$\frac{4}{8}$	We must also multiply the numerator by 4. $\frac{1}{2} = \frac{\_\_}{8}$ (4 × 1)

$\frac{1}{4}$

$\frac{1}{7}$

$\frac{2}{5}$

$\frac{2}{7}$

$\frac{3}{8}$

$\frac{4}{9}$

$\frac{7}{10}$

$\frac{5}{12}$

Write the fractions:

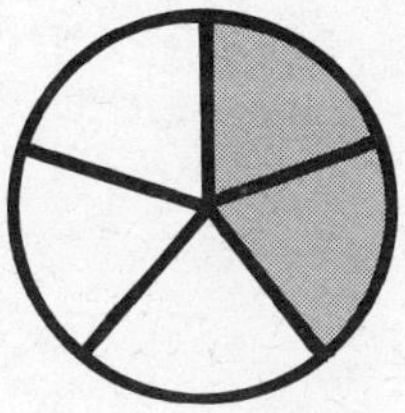

Answer	Frame
$\frac{5}{8}$	When we add or subtract fractions, we must have the same denominator: $\frac{2}{8} + \frac{3}{8} = \frac{}{8}$
	We add or subtract only the numerators. / denominator.
numerators	We can't add these fractions as they are. $\frac{1}{2} + \frac{1}{4}$ Can we add these fractions as they are? yes / no
$\frac{3}{4}$	But we know that $\frac{1}{2} = \frac{2}{4}$. So $\frac{2}{4} + \frac{1}{4} = \frac{}{4}$
4	The common denominator we found was 2. / 4.
yes	To make problems easy to work, we will try to find the lowest common denominator. This is the lowest number divisible by both denominators. Is 4 divisible by both 2 and 4? yes / no
yes	Is 4 the lowest common denominator for the fractions $\frac{3}{4}$ and $\frac{1}{2}$? yes / no
6	The lowest common denominator for the fractions $\frac{2}{3}$ and $\frac{1}{6}$ must be 3. / 6.

Adding Fractions

What will our denominator be? ____
(How many parts does the circle have?)

7

$\frac{1}{7} + \frac{1}{7} = \frac{2}{7}$

Did adding the fractions change the number of parts the circle has?
yes
no

no

We added only the numerators.

The denominator — stays the same. / changes.

stays the same

Remember to add only the numerators:

$\frac{1}{5} + \frac{1}{5} = \frac{\quad}{5}$

$\frac{2}{5}$

$\frac{1}{9} + \frac{1}{9} + \frac{2}{9} = \frac{\quad}{9}$

$\frac{4}{9}$

10	Find the smallest number divisible by each of the
21	following pairs:
$9 \times 4 = 36$	2 and 5 ($2 \times 5 =$ _____)
$5 \times 8 = 40$	3 and 7 ($3 \times 7 =$ _____)
$7 \times 8 = 56$	9 and 4 5 and 8 8 and 7
4	Since $4 \div 4 = 1$, the lowest number divisible by 4 is < 4 / 12
5	Since $5 \div 5 = 1$, the lowest number divisible by 5 is < 5 / 15
17	The lowest number divisible by 17 is ________ .
2	$4 \div 2 =$
yes	Is 4 divisible by both itself and 2 ? yes / no
4	The lowest number divisible by both 4 and 2 is____.
6	Find the lowest number divisible by each of the following pairs: 3 and 6
12	4 and 12
15	5 and 15

TEST

$\frac{2}{11}$	$\frac{1}{11} + \frac{1}{11} = \frac{\ \ }{11}$
$\frac{2}{15}$	$\frac{1}{15} + \frac{1}{15} = \frac{2}{\ \ }$
$\frac{2}{19}$	$\frac{1}{19} + \frac{1}{19} = \frac{\ \ }{19}$
$\frac{3}{4}$	$\frac{1}{4} + \frac{2}{4} = \frac{3}{\ \ }$
$\frac{3}{7}$	$\frac{1}{7} + \frac{2}{7} = \frac{\ \ }{7}$
$\frac{3}{8}$	$\frac{1}{8} + \frac{1}{8} + \frac{1}{8} = \frac{\ \ }{\ \ }$
$\frac{4}{5}$	$\frac{2}{5} + \frac{2}{5} = \frac{\ \ }{\ \ }$
$\frac{5}{7}$	$\frac{3}{7} + \frac{2}{7} = \frac{\ \ }{\ \ }$
$\frac{8}{9}$	$\frac{4}{9} + \frac{4}{9} = \frac{\ \ }{\ \ }$
$\frac{7}{11}$	$\frac{2}{11} + \frac{3}{11} + \frac{2}{11} = \frac{\ \ }{\ \ }$

3 2	Since 2 × 3 = 6, we know that 6 ÷ 2 = ____ and 6 ÷ 3 = ____ 6 is <u>divisible</u> by both 3 and 2.
20	To find the smallest number that is divisible by both 4 and 5, we say 4 × 5 =
12	To find the smallest number that is divisible by both 3 and 4, we say 3 × 4 =
30	Find the smallest number divisible by both 5 and 6: 5 × 6 =
7	Find the smallest number divisible by both 4 and 7: 4 × ____ = 28
7 × 5 = 35	Find the smallest number divisible by both 7 and 5: ____ × ____ = ____

$\frac{9}{16}$	$\frac{5}{16} + \frac{4}{16} = $ ___
$\frac{12}{23}$	$\frac{6}{23} + \frac{6}{23} = $ ___
$\frac{14}{15}$	$\frac{8}{15} + \frac{6}{15} = $ ___
$\frac{16}{31}$	$\frac{9}{31} + \frac{7}{31} = $ ___
$\frac{18}{19}$	$\frac{10}{19} + \frac{8}{19} = $ ___
$\frac{17}{26}$	$\frac{7}{26} + \frac{5}{26} + \frac{5}{26} = $ ___
$\frac{13}{18}$	$\frac{6}{18} + \frac{7}{18} = $ ___
$\frac{19}{35}$	$\frac{12}{35} + \frac{7}{35} = $ ___
$\frac{40}{67}$	$\frac{30}{67} + \frac{10}{67} = $ ___
$\frac{20}{103}$	$\frac{14}{103} + \frac{6}{103} = $ ___

TEST

Answers	Problems
$4\frac{1}{6}$, $11\frac{1}{7}$	$1\frac{5}{6} \times 2\frac{3}{11} = ?$ $5\frac{1}{7} \times 2\frac{1}{6} = ?$
$5\frac{2}{3}$, $4\frac{2}{5}$	$4\frac{1}{4} \times 1\frac{1}{3} = ?$ $3\frac{2}{3} \times 1\frac{1}{5} = ?$
$13\frac{1}{2}$, $14\frac{2}{3}$	$7\frac{1}{2} \times 1\frac{4}{5} = ?$ $6\frac{2}{3} \times 2\frac{1}{5} = ?$
$1\frac{7}{10}$, $3\frac{1}{11}$	$2\frac{1}{8} \div 1\frac{1}{4} = ?$ $6\frac{4}{5} \div 2\frac{1}{5} = ?$
$1\frac{1}{2}$, $1\frac{11}{14}$	$3\frac{3}{8} \div 2\frac{1}{4} = ?$ $5\frac{5}{7} \div 3\frac{1}{5} = ?$

Subtracting Fractions

Answers	Exercises
yes	$\frac{4}{5} - \frac{1}{5} = \frac{3}{5}$ Do we keep the same denominator? yes / no
$\frac{1}{3}$ $\frac{1}{4}$	Remember to subtract only the numerators: $\frac{2}{3} - \frac{1}{3} = \frac{\_\_}{3}$ $\frac{3}{4} - \frac{2}{4} = \frac{\_\_}{4}$
$\frac{2}{5}$ $\frac{3}{7}$	$\frac{4}{5} - \frac{2}{5} = \frac{\_\_}{5}$ $\frac{5}{7} - \frac{2}{7} = \frac{3}{\_\_}$
$\frac{2}{9}$ $\frac{1}{8}$	$\frac{7}{9} - \frac{5}{9} = \frac{\_\_}{9}$ $\frac{5}{8} - \frac{4}{8} = \frac{1}{\_\_}$
$\frac{1}{6}$ $\frac{3}{10}$	$\frac{4}{6} - \frac{3}{6} = \_\_$ $\frac{9}{10} - \frac{6}{10} = \_\_$

$\frac{11}{\cancel{2}_1} \times \frac{\cancel{4}^2}{7} = \frac{22}{7} = 3\frac{1}{7}$

$\frac{\cancel{24}^6}{5} \times \frac{3}{\cancel{4}_1} = \frac{18}{5} = 3\frac{3}{5}$

$\frac{13}{\cancel{3}_1} \times \frac{\cancel{3}^1}{4} = \frac{13}{4} = 3\frac{1}{4}$

$\frac{\cancel{15}^3}{\cancel{4}_2} \times \frac{\cancel{2}^1}{\cancel{5}_1} = \frac{3}{2} = 1\frac{1}{2}$

$\frac{\cancel{15}^3}{2} \times \frac{3}{\cancel{20}_4} = \frac{9}{8} = 1\frac{1}{8}$

$\frac{29}{\cancel{6}_2} \times \frac{\cancel{3}^1}{10} = \frac{29}{20} = 1\frac{9}{20}$

$\frac{11}{\cancel{2}_1} \times \frac{\cancel{2}^1}{3} = \frac{11}{3} = 3\frac{2}{3}$

$\frac{19}{\cancel{4}_1} \times \frac{\cancel{4}^1}{9} = \frac{19}{9} = 2\frac{1}{9}$

$5\frac{1}{2} \div 1\frac{3}{4} = ?$ $\quad 4\frac{4}{5} \div 1\frac{1}{3} = ?$

$4\frac{1}{3} \div 1\frac{1}{3} = ?$ $\quad 3\frac{3}{4} \div 2\frac{1}{2} = ?$

$7\frac{1}{2} \div 6\frac{2}{3} = ?$ $\quad 4\frac{5}{6} \div 3\frac{1}{3} = ?$

$5\frac{1}{2} \div 1\frac{1}{2} = ?$ $\quad 4\frac{3}{4} \div 2\frac{1}{4} = ?$

$\frac{3}{16}$	$\frac{9}{16} - \frac{6}{16} =$ ___
$\frac{7}{12}$	$\frac{11}{12} - \frac{4}{12} =$ ___
$\frac{4}{15}$	$\frac{7}{15} - \frac{3}{15} =$ ___
$\frac{14}{27}$	$\frac{19}{27} - \frac{5}{27} =$ ___
$\frac{5}{18}$	$\frac{8}{18} - \frac{3}{18} =$ ___
$\frac{1}{32}$	$\frac{11}{32} - \frac{10}{32} =$ ___
$\frac{7}{19}$	$\frac{14}{19} - \frac{7}{19} =$ ___
$\frac{16}{43}$	$\frac{26}{43} - \frac{10}{43} =$ ___
$\frac{41}{68}$	$\frac{47}{68} - \frac{6}{68} =$ ___
$\frac{3}{107}$	$\frac{23}{107} - \frac{20}{107} =$ ___

$\frac{11}{4}$	$2\frac{3}{4} \div 1\frac{1}{4} = ?$ $2\frac{3}{4} =$
$\frac{5}{4}$	$1\frac{1}{4} =$
$\frac{4}{5}$	$\frac{11}{4} \div \frac{5}{4} = ?$ $\frac{5}{4}$ inverted is ——.
$\frac{11}{5}$	$\frac{11}{\cancel{4}_{1}} \times \frac{\cancel{4}^{1}}{5} =$
$2\frac{1}{5}$	$\frac{11}{5} =$
$\frac{11}{3}$	$3\frac{2}{3} \div 2\frac{3}{4} = ?$ $3\frac{2}{3} =$
$\frac{11}{4}$	$2\frac{3}{4} =$
$\frac{4}{3}$	$\frac{11}{3} \div \frac{11}{4} = ?$ $\frac{\cancel{11}^{1}}{3} \times \frac{4}{\cancel{11}_{1}} =$
$1\frac{1}{3}$	$\frac{4}{3} =$

Multiplying Fractions

Answers	
6	Here is a bundle of sticks.  How many sticks are there? ____ This stick is one of six. "one of six" → $\frac{1}{\ }$
$\frac{1}{6}$	So this stick is $\frac{1}{6}$ of the whole bundle.
$\frac{1}{2}$	We split the bundle into two parts.  This part is one of two. "one of two" → $\frac{\ }{2}$ So this part is $\frac{1}{2}$ of the whole bundle.
3	How many sticks are there in $\frac{1}{2}$ of the bundle? ____
$\frac{1}{3}$	 This stick is one of three sticks in $\frac{1}{2}$ of the bundle. "one of three" → $\frac{1}{\ }$
$\frac{1}{6}$	So this stick is $\frac{1}{3}$ of $\frac{1}{2}$ of the bundle. It is also $\frac{\ }{6}$ of the whole bundle.
$\frac{1}{2}$	So $\frac{1}{3}$ of $\frac{1}{\ }$ is the same as $\frac{1}{6}$.
$\frac{1}{6}$	Here is a short way to get the same answer. Multiply numerator × numerator and denominator × denominator $\frac{1 \times 1}{3 \times 2} = \frac{\ }{6}$

$\frac{13}{\cancel{3}_1} \times \frac{\cancel{9}^3}{2} = \frac{39}{2} = 19\frac{1}{2}$

$\frac{\cancel{21}^3}{4} \times \frac{9}{\cancel{7}_1} = \frac{27}{4} = 6\frac{3}{4}$

$\frac{7}{4} \times \frac{5}{2} = \frac{35}{8} = 4\frac{3}{8}$

$\frac{15}{\cancel{7}_1} \times \frac{\cancel{7}^1}{4} = \frac{15}{4} = 3\frac{3}{4}$

$\frac{\cancel{24}^3}{7} \times \frac{13}{\cancel{8}_1} = \frac{39}{7} = 5\frac{4}{7}$

$\frac{7}{\cancel{3}_1} \times \frac{\cancel{6}^2}{5} = \frac{14}{5} = 2\frac{4}{5}$

$\frac{\cancel{5}^1}{2} \times \frac{11}{\cancel{10}_2} = \frac{11}{4} = 2\frac{3}{4}$

$\frac{11}{\cancel{5}_1} \times \frac{\cancel{10}^2}{3} = \frac{22}{3} = 7\frac{1}{3}$

$\frac{7}{\cancel{2}_1} \times \frac{\cancel{10}^5}{3} = \frac{35}{3} = 11\frac{2}{3}$

$\frac{19}{\cancel{7}_1} \times \frac{\cancel{14}^2}{5} = \frac{38}{5} = 7\frac{3}{5}$

$4\frac{1}{3} \times 4\frac{1}{2} = ?$ $\qquad$ $5\frac{1}{4} \times 1\frac{2}{7} = ?$

$1\frac{3}{4} \times 2\frac{1}{2} = ?$ $\qquad$ $2\frac{1}{7} \times 1\frac{3}{4} = ?$

$3\frac{3}{7} \times 1\frac{5}{8} = ?$ $\qquad$ $2\frac{1}{3} \times 1\frac{1}{5} = ?$

$2\frac{1}{2} \times 1\frac{1}{10} = ?$ $\qquad$ $2\frac{1}{5} \times 3\frac{1}{3} = ?$

$3\frac{1}{2} \times 3\frac{1}{3} = ?$ $\qquad$ $2\frac{5}{7} \times 2\frac{4}{5} = ?$

$\frac{1}{2}$

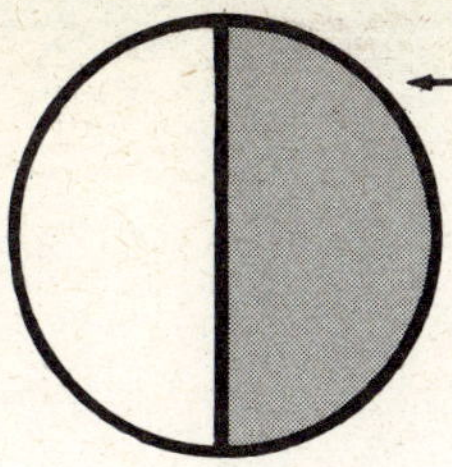

Here is one of two parts of a circle.

It is $\frac{1}{2}$ / $\frac{1}{3}$ of the circle.

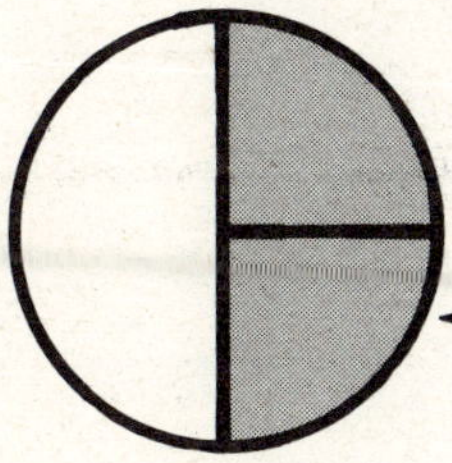

Here is the same $\frac{1}{2}$ cut into two parts.

$\frac{1}{2}$

This is one of two parts of $\frac{1}{2}$ of a circle.

It is $\frac{1}{2}$ of $\frac{1}{}$.

$\frac{1}{4}$

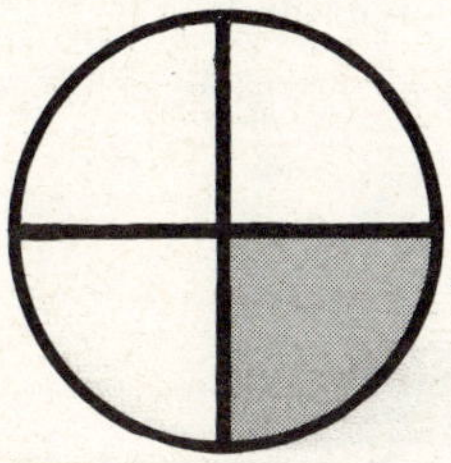

Now the circle has four parts.

This is one of four parts.

It is $\frac{1}{}$ of the whole circle.

---

yes

Is it also $\frac{1}{2}$ of $\frac{1}{2}$ of the circle? yes / no

---

$\frac{1}{2}$

So $\frac{1}{2}$ of $\frac{1}{}$ is the same as $\frac{1}{4}$.

$\frac{1}{4}$

We could have multiplied

numerator × numerator
and denominator × denominator

$$\frac{1 \times 1}{2 \times 2} = \frac{1}{}$$

---

c.

We multiplied: (Check the correct answer.)

___ a. only the numerators.
___ b. only the denominators.
___ c. both the numerators and the denominators.

	$1\frac{3}{5} \times 4\frac{1}{2} = ?$
$\frac{8}{5}$	$1\frac{3}{5} =$
$\frac{9}{2}$	$4\frac{1}{2} =$
$\frac{8}{5} \times \frac{9}{2} = \frac{36}{5}$	$\frac{8}{5} \times \frac{9}{2} =$
$7\frac{1}{5}$	$\frac{36}{5} =$
	$2\frac{2}{3} \times 3\frac{1}{4} = ?$
$\frac{8}{3}$	$2\frac{2}{3} =$
$\frac{13}{4}$	$3\frac{1}{4} =$
$\frac{8}{3} \times \frac{13}{4} = \frac{26}{3}$	$\frac{8}{3} \times \frac{13}{4} =$
$8\frac{2}{3}$	$\frac{26}{3} =$
	$3\frac{3}{4} \times 2\frac{1}{5} = ?$
	$\downarrow \quad \downarrow$
$\frac{15}{4} \times \frac{11}{5} = \frac{33}{4}$	$\frac{15}{4} \times \frac{11}{5} =$
$8\frac{1}{4}$	$\frac{33}{4} =$

TEST

$\frac{4}{5}$	$\frac{1}{5} + \frac{3}{5} =$ ___
$\frac{3}{11}$	$\frac{2}{11} + \frac{1}{11} =$ ___
$\frac{8}{9}$	$\frac{5}{9} + \frac{3}{9} =$ ___
$\frac{6}{17}$	$\frac{1}{17} + \frac{5}{17} =$ ___
$\frac{16}{25}$	$\frac{13}{25} + \frac{3}{25} =$ ___
$\frac{2}{7}$	$\frac{5}{7} - \frac{3}{7} =$ ___
$\frac{5}{9}$	$\frac{8}{9} - \frac{3}{9} =$ ___
$\frac{1}{6}$	$\frac{5}{6} - \frac{4}{6} =$ ___
$\frac{4}{13}$	$\frac{11}{13} - \frac{7}{13} =$ ___
$\frac{33}{74}$	$\frac{39}{74} - \frac{6}{74} =$ ___

$$2\frac{1}{3} \times 1\frac{1}{2} = ?$$

$\frac{7}{3}$

First change the mixed numbers to improper fractions.

$$2\frac{1}{3} = \_\_$$

$\frac{3}{2}$

$$1\frac{1}{2} = \_\_$$

Multiply:

$\frac{7}{2}$

$$\frac{7}{\cancel{3}_{1}} \times \frac{\cancel{3}^{1}}{2} = \_\_$$

Change $\frac{7}{2}$ to a mixed number.

$3\frac{1}{2}$

$$\frac{7}{2} = 3\frac{1}{\_\_}$$

$\frac{10}{3}$

$$3\frac{1}{3} \times 2\frac{1}{4} = ?$$

$\frac{9}{4}$

$$3\frac{1}{3} = \_\_$$

$\frac{15}{2}$

$$2\frac{1}{4} = \_\_$$

$7\frac{1}{2}$

So $$\frac{\cancel{10}^{5}}{\cancel{3}_{1}} \times \frac{\cancel{9}^{3}}{\cancel{4}_{2}} = \frac{15}{\_\_}$$

$$\frac{15}{2} =$$

$$\frac{1}{2} \text{ of } \frac{1}{2} = \frac{?}{?}$$

We find the answer by multiplying:

$$\frac{1 \times 1}{2 \times 2} = \frac{1}{}$$

Answer: $\frac{1}{4}$

$$\frac{1}{3} \text{ of } \frac{1}{3} = \frac{?}{?}$$

Multiply:

$$\frac{1 \times 1}{3 \times 3} = \frac{}{}$$

Answer: $\frac{1}{9}$

We write only one multiplication sign:

$$\frac{2}{3} \times \frac{1}{2} \quad \text{means} \quad \frac{2 \times 1}{3 \times 2}$$

Multiply

$$\frac{2}{3} \times \frac{1}{2} = \frac{}{6}$$

Answer: $\frac{2}{6}$

$$\frac{1}{2} \times \frac{3}{4} = \frac{3}{}$$

Answer: $\frac{3}{8}$

$$\frac{3}{5} \times \frac{1}{2} = \frac{}{}$$

Answer: $\frac{3}{10}$

$$\frac{2}{3} \times \frac{1}{3} = \frac{}{}$$

Answer: $\frac{2}{9}$

TEST

Answers		Problems	
$\frac{5}{2}$	$\frac{4}{3}$	$2\frac{1}{2} = \frac{5}{}$	$1\frac{1}{3} = \frac{4}{}$
$\frac{7}{2}$	$\frac{12}{5}$	$3\frac{1}{2} = \frac{}{2}$	$2\frac{2}{5} = \frac{}{5}$
$\frac{30}{7}$	$\frac{15}{4}$	$4\frac{2}{7} = \frac{30}{}$	$3\frac{3}{4} = \frac{}{4}$
$\frac{17}{6}$	$\frac{15}{8}$	$2\frac{5}{6} = \frac{}{6}$	$1\frac{7}{8} = \frac{15}{}$
$\frac{14}{5}$	$\frac{19}{3}$	$2\frac{4}{5} = $ ___	$6\frac{1}{3} = $ ___

$\frac{1}{12}$	$\frac{1}{3} \times \frac{1}{4} =$ ___
$\frac{1}{10}$	$\frac{1}{2} \times \frac{1}{5} =$ ___
$\frac{3}{10}$	$\frac{3}{5} \times \frac{1}{2} =$ ___
$\frac{1}{15}$	$\frac{1}{3} \times \frac{1}{5} =$ ___
$\frac{2}{15}$	$\frac{1}{5} \times \frac{2}{3} =$ ___
$\frac{1}{12}$	$\frac{1}{2} \times \frac{1}{6} =$ ___
$\frac{4}{15}$	$\frac{1}{3} \times \frac{4}{5} =$ ___
$\frac{5}{12}$	$\frac{5}{6} \times \frac{1}{2} =$ ___
$\frac{1}{18}$	$\frac{1}{3} \times \frac{1}{6} =$ ___
$\frac{3}{20}$	$\frac{1}{5} \times \frac{3}{4} =$ ___

an improper	We can change a mixed number to < a proper / an improper fraction.
4	$4\frac{2}{3} = ?$
2	$3 \times \_\_\_ = 12$ $12 + \_\_\_ = 14$
$\frac{14}{3}$	So $4\frac{2}{3} = \frac{\quad}{3}$
5	$3\frac{4}{5} = ?$ $\_\_\_ \times 3 = 15$
15	$\_\_\_ + 4 = 19$
$\frac{19}{5}$	So $3\frac{4}{5} = \frac{19}{\quad}$
$6 \times 4 = 24$	$4\frac{5}{6} = ?$ $6 \times \_\_\_ = \_\_\_$
29	$24 + 5 = \_\_\_$
$\frac{29}{6}$	So $4\frac{5}{6} = \frac{\quad}{\quad}$

	$\frac{3}{5}$ is a <u>proper</u> fraction.
smaller	Its numerator is ⟨ larger / smaller ⟩ than its denominator.
	$\frac{9}{2}$ is an <u>improper</u> fraction.
yes	Is its numerator larger than its denominator? yes / no
$\frac{7}{4}$ $\frac{12}{5}$ $\frac{8}{3}$	Circle the improper fractions: (circled) $\frac{3}{2}$ $\frac{1}{8}$ $\frac{7}{4}$ $\frac{12}{5}$ $\frac{3}{14}$ $\frac{11}{19}$ $\frac{8}{3}$
proper	The numbers you did not circle are called ⟨ proper / improper ⟩ fractions.
	To invert a fraction is to turn it over. $\frac{2}{5}$ inverted is $\frac{5}{2}$
$\frac{7}{3}$	$\frac{3}{7}$ inverted is $\frac{7}{\_\_}$
$\frac{4}{9}$	$\frac{9}{4}$ inverted is $\frac{\_\_}{9}$
$\frac{3}{2}$	$\frac{2}{3}$ inverted is $\frac{\_\_}{\_\_}$
$\frac{6}{11}$	$\frac{11}{6}$ inverted is $\frac{\_\_}{\_\_}$

Answer	Frame
	Any mixed number may be changed to an improper fraction. $$2\frac{1}{3} = ?$$
6	Multiply the whole number by the denominator of the fraction: $$3 \times 2 = \_\_\_\_$$
7	Add the numerator of the fraction to this answer: $$6 + 1 = \_\_\_\_$$ This is the new numerator. The denominator stays the same: $2\frac{1}{3} = \frac{7}{3}$
12	$$3\frac{1}{4} = ?$$ $$4 \times 3 = \_\_\_\_$$
13	$$12 + 1 = \_\_\_\_$$ So $3\frac{1}{4} = \frac{13}{4}$
8	$$2\frac{3}{4} = ?$$
11	$$4 \times 2 = \_\_\_\_$$ $$8 + 3 = \_\_\_\_$$
$\frac{11}{4}$	So $2\frac{3}{4} = \frac{11}{\_\_\_\_}$

Answers	Frames
$\frac{8}{5}$ $\frac{7}{12}$ $\frac{19}{6}$	Invert these fractions: $\frac{5}{8}$ $\frac{12}{7}$ $\frac{6}{19}$
$\frac{1}{5}$	If we invert the improper fraction $\frac{5}{1}$, we get the proper fraction —.
$\frac{6}{1}$	Any whole number can be thought of as an improper fraction with a denominator of 1. (Box: $7 \longrightarrow \frac{7}{1}$; $2 \longrightarrow \frac{2}{1}$) $6 \longrightarrow \frac{\ }{1}$
$\frac{4}{1}$	$4 \longrightarrow$ —
$\frac{17}{1}$	$17 \longrightarrow$ —
$\frac{65}{1}$	$65 \longrightarrow$ —
$\frac{7}{1}$	The whole number 7 can be thought of as $\frac{\ }{1}$. So if we invert the whole number 7,
$\frac{1}{7}$	we get the proper fraction $\frac{1}{\ }$.
$\frac{1}{9}$ $\frac{1}{48}$ $\frac{1}{3}$ $\frac{1}{179}$	Invert: $\frac{9}{1}$ $\frac{48}{1}$ 3 179

$\frac{3}{8}$	$1 - \frac{5}{8}$	$3 - \frac{5}{9}$
$2\frac{4}{9}$		
$5\frac{1}{3}$	$6 - \frac{2}{3}$	$11 - \frac{5}{7}$
$10\frac{2}{7}$		
$16\frac{1}{9}$	$17 - \frac{8}{9}$	$26 - \frac{1}{4}$
$25\frac{3}{4}$		
$35\frac{2}{5}$	$39 - 3\frac{3}{5}$	$154\frac{2}{14} - 2\frac{5}{14}$
$151\frac{11}{14}$		
$367\frac{5}{9}$	$689 - 321\frac{4}{9}$	$1,321\frac{5}{11} - 200\frac{6}{11}$
$1,120\frac{10}{11}$		

3	$\frac{4}{2}$ ← This line means <u>divide</u> <u>if</u> <u>possible</u>. $\frac{4}{2} = 2 \quad \left(2\overline{)\,4}\right)$ with quotient 2 We divided the numerator by the denominator.
4 whole numbers	$\frac{6}{2} =$ $\qquad$ $\frac{8}{2} =$ We changed the improper fractions to — proper fractions. / whole numbers.
2 3	$\frac{10}{5} =$ $\qquad$ $\frac{9}{3} =$
6 5	$\frac{12}{2} =$ $\qquad$ $\frac{15}{3} =$
3 4	$\frac{21}{7} =$ $\qquad$ $\frac{16}{4} =$
7 6	$\frac{35}{5} =$ $\qquad$ $\frac{24}{4} =$
9 8	$\frac{63}{7} =$ $\qquad$ $\frac{72}{9} =$

$7\frac{3}{14}$	$\begin{array}{r} 9 \\ -\ 1\frac{11}{14} \\ \hline ? \end{array}$	⟶	$\begin{array}{r} 8\frac{14}{14} \\ -\ 1\frac{11}{14} \\ \hline 7\frac{\ }{14} \end{array}$
$100\frac{5}{9}$	$\begin{array}{r} 110\frac{3}{9} \\ -\ \ 9\frac{7}{9} \\ \hline ? \end{array}$	⟶	$\begin{array}{r} 109\frac{12}{9} \\ -\ \ 9\frac{7}{9} \\ \hline 100\ — \end{array}$
$\begin{array}{r} 24\frac{10}{8}\ \left(\frac{2}{8}+\frac{8}{8}\right) \\ -\ 15\frac{7}{8} \\ \hline 9\frac{3}{8} \end{array}$	$\begin{array}{r} 25\frac{2}{8} \\ -\ 15\frac{7}{8} \\ \hline ? \end{array}$	⟶	
$\begin{array}{r} 96\frac{5}{5} \\ -\ \ 4\frac{4}{5} \\ \hline 92\frac{1}{5} \end{array}$	$\begin{array}{r} 97 \\ -\ \ 4\frac{4}{5} \\ \hline ? \end{array}$	⟶	
$\begin{array}{r} 87\frac{9}{9} \\ -\ 43\frac{2}{9} \\ \hline 44\frac{7}{9} \end{array}$	$\begin{array}{r} 88 \\ -\ 43\frac{2}{9} \\ \hline ? \end{array}$	⟶	
$\begin{array}{r} 246\frac{12}{11}\ \left(\frac{1}{11}+\frac{11}{11}\right) \\ -\ 113\frac{4}{11} \\ \hline 133\frac{8}{11} \end{array}$	$\begin{array}{r} 247\frac{1}{11} \\ -\ 113\frac{4}{11} \\ \hline ? \end{array}$	⟶	

Answers	TEST
$\frac{1}{8}$	$\frac{1}{2} \times \frac{1}{4} =$
$\frac{2}{15}$	$\frac{2}{5} \times \frac{1}{3} =$
$\frac{8}{21}$	$\frac{4}{7} \times \frac{2}{3} =$
$\frac{5}{54}$	$\frac{5}{9} \times \frac{1}{6} =$
$\frac{7}{2}$	Circle the improper fraction: $\frac{2}{9}$ $\frac{7}{2}$
$\frac{5}{9}$ $\frac{12}{7}$	Invert these fractions: $\frac{9}{5}$ $\frac{7}{12}$
2	Change the improper fractions to whole numbers: $\frac{6}{3} =$
5	$\frac{20}{4} =$
9	$\frac{45}{5} =$

$$41 - \frac{5}{13} = \;?$$

$$41 = 40\frac{13}{}$$

$$40\frac{13}{13} - \frac{5}{13}$$

$$52 - \frac{5}{6} = \;?$$

$$113 - \frac{4}{11} = \;?$$

$$18 - \frac{3}{4} = \;?$$

$$73 - \frac{7}{9} = \;?$$

$$40\frac{13}{13}$$

$$40\frac{8}{13}$$

$$51\frac{6}{6} - \frac{5}{6} = 51\frac{1}{6}$$

$$112\frac{11}{11} - \frac{4}{11} = 112\frac{7}{11}$$

$$17\frac{4}{4} - \frac{3}{4} = 17\frac{1}{4}$$

$$72\frac{9}{9} - \frac{7}{9} = 72\frac{2}{9}$$

Dividing Fractions

	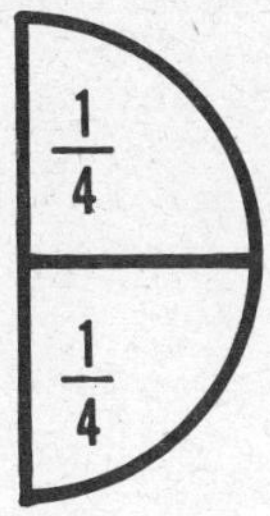 Here is $\frac{1}{2}$ of a circle.
	How many $\frac{1}{4}$'s are there in
2	$\frac{1}{2}$ of a circle? ______
	So $\frac{1}{2} \div \frac{1}{4} = 2$
	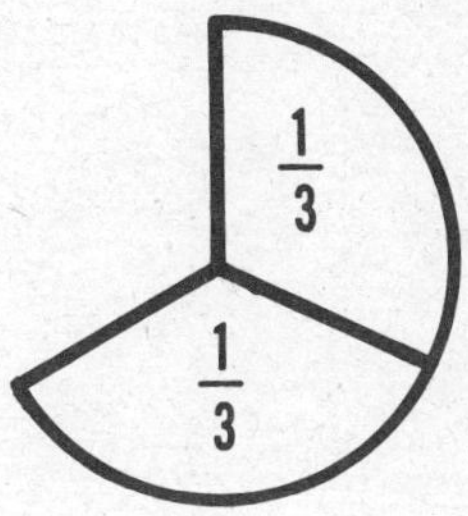 Here is $\frac{2}{3}$ of a circle.
	How many $\frac{1}{3}$'s are there in
2	$\frac{2}{3}$ of a circle? ____
2	So $\frac{2}{3} \div \frac{1}{3} =$
	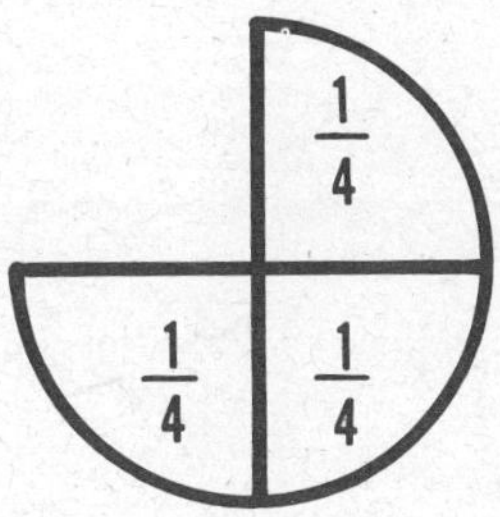 Here is $\frac{3}{4}$ of a circle.
	How many $\frac{1}{4}$'s are there in
3	$\frac{3}{4}$? ______
3	So $\frac{3}{4} \div \frac{1}{4} =$
	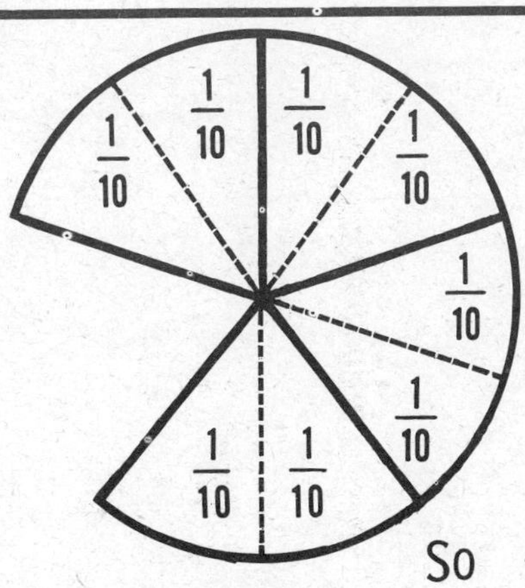 Here is $\frac{4}{5}$ of a circle.
	How many $\frac{1}{10}$'s are there in
8	$\frac{4}{5}$? ______
8	So $\frac{4}{5} \div \frac{1}{10} =$

$3\frac{1}{5}$	**TEST**
	$3 + \frac{1}{5} =$
$19\frac{3}{4}$	$19 + \frac{3}{4} =$
$8\frac{5}{8}$	$6 + 2 + \frac{5}{8}$
$16\frac{9}{10}$	$7 + 5 + 4 + \frac{9}{10}$
$3\frac{1}{6}$	$4\frac{1}{6} - 1$
$5\frac{4}{7}$	$18\frac{4}{7} - 13$
$\frac{3}{5}$	$1 - \frac{2}{5} =$
$\frac{5}{8}$	$1 - \frac{3}{8} =$
$3\frac{2}{3}$	$4 - \frac{1}{3}$
$14\frac{7}{9}$	$15 - \frac{2}{9}$

$$\frac{1}{2} \div \frac{1}{6} = ?$$

$\frac{1}{6}$

We are dividing $\frac{1}{2}$ by $\frac{1}{\ }$.

To divide by a fraction, we must first invert it.

$\frac{6}{1}$

$\frac{1}{6}$ inverted is ___.

$\frac{6}{2}$

Now we multiply:

$$\frac{1}{2} \times \frac{6}{1} = \frac{\ }{2}$$

3

But $\frac{6}{2}$ is an improper fraction.

$$\frac{6}{2} =$$

So there are three $\frac{1}{6}$'s in $\frac{1}{2}$.

Let's check with a circle.

We divide the circle into six parts.

Each part is $\frac{1}{6}$ of the circle.

How many of these parts are there in $\frac{1}{2}$ of the circle? ___

3

---

correct

Our answer is correct. / incorrect.

$8\frac{5}{7}$	$\begin{array}{r} 9 \\ -\ \frac{2}{7} \\ \hline ? \end{array}$ Change the 9 to $8\frac{7}{7}$. $\begin{array}{r} 8\frac{7}{7} \\ -\ \frac{2}{7} \\ \hline 8\ \text{—} \end{array}$
$6\frac{8}{8}$ $6\frac{3}{8}$	$\begin{array}{r} 7 \\ -\ \frac{5}{8} \\ \hline ? \end{array}$ Change the 7 to $6\frac{}{8}$. $\begin{array}{r} 6\frac{8}{8} \\ -\ \frac{5}{8} \\ \hline 6\ \text{—} \end{array}$
$11\frac{9}{9}$ $11\frac{7}{9}$	$\begin{array}{r} 12 \\ -\ \frac{2}{9} \\ \hline ? \end{array}$ $12 = 11\frac{}{9}$ $\begin{array}{r} 11\frac{9}{9} \\ -\ \frac{2}{9} \\ \hline \end{array}$

$\frac{8}{1}$	$\frac{1}{2} \div \frac{1}{8} = ?$ $\frac{1}{8}$ inverted is ___.
$\frac{8}{2}$	$\frac{1}{2} \times \frac{8}{1} =$ ___
4	$\frac{8}{2} =$
multiply	To divide by a fraction, we invert it and then (multiply. / subtract.)
an improper	If we get (a proper / an improper) fraction,
b.	we then divide ___ a. the denominator by the numerator. ___ b. the numerator by the denominator.
$\frac{6}{1}$	$\frac{1}{3} \div \frac{1}{6} = ?$ $\frac{1}{6}$ inverted is ___.
$\frac{6}{3}$	$\frac{1}{3} \times \frac{6}{1} =$ ___
2	$\frac{6}{3} =$

$\frac{5}{5}$	$1 - \frac{3}{5} = ?$ $1 = \frac{}{5}$
$\frac{2}{5}$	$\frac{5}{5} - \frac{3}{5} = \frac{2}{}$
$\frac{6}{6}$	$\begin{array}{r} 1 \\ - \frac{5}{6} \\ \hline ? \end{array}$ $1 = \frac{}{6}$
$\frac{1}{6}$	$\frac{6}{6} - \frac{5}{6} = \_\_\_$
2	$\begin{array}{r} 2 \\ - \frac{1}{4} \\ \hline ? \end{array}$ How many 1's are there in 2 ? _____ Change one of them to $\frac{4}{4}$. $2 = 1\frac{4}{4}$
$1\frac{3}{4}$	$\begin{array}{r} 1\frac{4}{4} \\ - \frac{1}{4} \\ \hline 1\frac{}{4} \end{array}$

	$\frac{3}{4} \div \frac{4}{5} = ?$
$\frac{5}{4}$	$\frac{4}{5}$ inverted is —.
$\frac{15}{16}$	$\frac{3}{4} \times \frac{5}{4} = $ —
a proper	Our answer is a proper / an improper fraction.
	$\frac{2}{5} \div \frac{2}{15} = ?$
$\frac{15}{2}$	$\frac{2}{15}$ inverted is —.
$\frac{30}{10}$	$\frac{2}{5} \times \frac{15}{2} = $ —
3	$\frac{30}{10} =$
	$\frac{2}{3} \div \frac{1}{9} = ?$
$\frac{2}{3} \times \frac{9}{1} = \frac{18}{3}$	$\frac{2}{3} \times \frac{—}{1} = $ —
6	$\frac{18}{3} =$
	$\frac{2}{3} \div \frac{5}{7} = ?$
$\frac{2}{3} \times \frac{7}{5} = \frac{14}{15}$	$\frac{2}{3} \times$ — = —

1	$\frac{2}{2} = 1$ $\frac{6}{6} =$
1	$\frac{12}{12} =$
1	$\frac{203}{203} =$
$\frac{4}{4}$	$1 = \frac{3}{3}$ $1 = \frac{\quad}{4}$
$\frac{17}{17}$	$1 = \frac{17}{\quad}$
$\frac{305}{305}$	$1 = \frac{305}{\quad}$
$\frac{8}{8}$	$1 - \frac{3}{8} = ?$ $1 = \frac{\quad}{8}$
$\frac{5}{8}$	$\frac{8}{8} - \frac{3}{8} = \frac{\quad}{8}$

$$\frac{3}{4} \times \frac{8}{3} = \frac{24}{12} = 2$$

$$\frac{3}{4} \div \frac{3}{8} = ?$$

$$\frac{3}{4} \times \text{——} = \text{——} = 2$$

$$\frac{2}{5} \times \frac{10}{1} = \frac{20}{5} = 4$$

$$\frac{2}{5} \div \frac{1}{10} = ?$$

$$\times \frac{12}{1} = \frac{60}{6} = 10$$

$$\frac{5}{6} \div \frac{1}{12} = ?$$

$$\frac{4}{5} \times \frac{15}{2} = \frac{60}{10} = 6$$

$$\frac{4}{5} \div \frac{2}{15} = ?$$

$$\frac{5}{6} \times \frac{7}{6} = \frac{35}{36}$$

$$\frac{5}{6} \div \frac{6}{7} = ?$$

Answers	Problems	
$1\frac{1}{5}$ $2\frac{3}{8}$	Just bring down the fraction. $\begin{array}{r} 3\frac{1}{5} \\ -\ 2 \\ \hline \_\_\ \frac{1}{5} \end{array}$	$\begin{array}{r} 7\frac{3}{8} \\ -\ 5 \\ \hline 2\ \frac{3}{\_\_} \end{array}$
$4\frac{2}{3}$ $11\frac{5}{6}$	$\begin{array}{r} 10\frac{2}{3} \\ -\ 6 \\ \hline 4\ \_\_ \end{array}$	$\begin{array}{r} 15\frac{5}{6} \\ -\ 4 \\ \hline \end{array}$
$22\frac{7}{8}$ $25\frac{1}{2}$	$\begin{array}{r} 29\frac{7}{8} \\ -\ 7 \\ \hline \end{array}$	$\begin{array}{r} 45\frac{1}{2} \\ -20 \\ \hline \end{array}$
$43\frac{19}{27}$ $11\frac{74}{103}$	$\begin{array}{r} 97\frac{19}{27} \\ -\ 54 \\ \hline \end{array}$	$\begin{array}{r} 214\frac{74}{103} \\ -\ 203 \\ \hline \end{array}$
$29\frac{3}{4}$ $205\frac{3}{7}$	$\begin{array}{r} 78\frac{3}{4} \\ -\ 49 \\ \hline \end{array}$	$\begin{array}{r} 1,463\frac{3}{7} \\ -\ 1,258 \\ \hline \end{array}$

$\frac{9}{11}$ $\frac{3}{17}$	**TEST** $\frac{7}{11} + \frac{2}{11} =$ $\frac{8}{17} - \frac{5}{17} =$
$\frac{4}{15}$ $\frac{6}{21}$	$\frac{2}{5} \times \frac{2}{3} =$ $\frac{3}{7} \times \frac{2}{3} =$
4 2	$\frac{2}{3} \div \frac{1}{6} = ?$ $\frac{3}{4} \div \frac{3}{8} = ?$
3 $\frac{14}{15}$	$\frac{1}{2} \div \frac{1}{6} = ?$ $\frac{2}{3} \div \frac{5}{7} = ?$
3 $\frac{24}{25}$	$\frac{1}{3} \div \frac{1}{9} = ?$ $\frac{4}{5} \div \frac{5}{6} = ?$

Answers	Problems
	$2 + \frac{1}{2} = 2\frac{1}{2}$
$4\frac{1}{3}$	$4 + \frac{1}{3} = 4\frac{}{3}$
$7\frac{2}{5}$	$7 + \frac{2}{5} = 7$ —
$3\frac{4}{9}$	$3 + \frac{4}{9} =$
$11\frac{7}{8}$	$11 + \frac{7}{8} =$
$7\frac{2}{3}$	$3 + 4 + \frac{2}{3} =$ — $\frac{2}{3}$
$11\frac{3}{8}$	$5 + 6 + \frac{3}{8} = 11$ —
$14\frac{1}{4}$	$5 + 7 + 2 + \frac{1}{4} =$
$33\frac{6}{11}$	$10 + 9 + 14 + \frac{6}{11} =$
$16\frac{6}{7}$	$7 + 4 + 5 + \frac{6}{7} =$
$23\frac{7}{10}$	$9 + 6 + 8 + \frac{7}{10} =$

	To divide or multiply a fraction by a whole number, think of the whole number as an improper fraction. $\frac{3}{1} \times \frac{1}{4} = \frac{3}{4}$
$\frac{2}{3}$	$\frac{2}{1} \times \frac{1}{3} = \frac{\quad}{3}$
$\frac{4}{7}$	$4 \times \frac{1}{7} = \frac{4}{\quad}$
$\frac{4}{5}$	$2 \times \frac{2}{5} = \text{—}$
$\frac{6}{11}$	$3 \times \frac{2}{11} = \text{—}$
$\frac{1}{6}$	$\frac{1}{3} \div \frac{2}{1} = ?$ Invert the improper fraction and multiply: $\frac{1}{3} \times \frac{1}{2} = \frac{1}{\quad}$
$\frac{1}{4} \times \frac{1}{3} = \frac{1}{12}$	$\frac{1}{4} \div \frac{3}{1} = ?$ $\frac{1}{4} \times \frac{1}{\quad} = \frac{1}{12}$

$$\frac{12}{8} = \frac{3}{2} = 1\frac{1}{2}$$

$$\frac{28}{21} = \frac{4}{3} = 1\frac{1}{3}$$

$$\frac{14}{10} = \frac{7}{5} = 1\frac{2}{5}$$

$$\frac{12}{9} = \frac{4}{3} = 1\frac{1}{3}$$

$$\frac{9}{\cancel{4}_{1}} \times \frac{\cancel{16}^{4}}{5} = \frac{36}{5} = 7\frac{1}{5}$$

$$\frac{\cancel{18}^{6}}{7} \times \frac{10}{\cancel{3}_{1}} = \frac{60}{7} = 8\frac{4}{7}$$

$$\frac{13}{\cancel{14}_{2}} \times \frac{\cancel{7}^{1}}{3} = \frac{13}{6} = 2\frac{1}{6}$$

$$\frac{\cancel{10}^{2}}{5} \times \frac{7}{\cancel{8}_{1}} = \frac{14}{5} = 2\frac{4}{5}$$

$$\frac{7}{\cancel{12}_{3}} \times \frac{\cancel{8}^{2}}{3} = \frac{14}{9} = 1\frac{5}{9}$$

$$\frac{\cancel{20}^{5}}{3} \times \frac{1}{\cancel{4}_{1}} = \frac{5}{3} = 1\frac{2}{3}$$

TEST

$$\frac{7}{8} + \frac{5}{8} = \text{—} = \text{—} =$$

$$\frac{17}{21} + \frac{11}{21} = \text{—} = \text{—} =$$

$$\frac{17}{10} - \frac{3}{10} = \text{—} = \text{—} =$$

$$\frac{25}{9} - \frac{13}{9} = \text{—} = \text{—} =$$

$$\frac{9}{4} \times \frac{16}{5} = \text{—} =$$

$$\frac{18}{7} \times \frac{10}{3} = \text{—} =$$

$$\frac{16}{5} \div \frac{8}{7} = ?$$

$$\frac{13}{14} \times \frac{7}{3} = \text{—} =$$

$$\frac{7}{12} \div \frac{3}{8} = ?$$

$$\frac{20}{3} \div 4 = ?$$

$\frac{1}{4}$	$\frac{3}{7} \div 4 = ?$
$\frac{3}{7} \times \frac{1}{4} = \frac{3}{28}$	4 inverted is ____ . $\frac{3}{7} \times$ ____ $=$ ____
$\frac{2}{3} \times \frac{1}{5} = \frac{2}{15}$	$\frac{2}{3} \div 5 - ?$
$\frac{5}{6} \times \frac{1}{3} = \frac{5}{18}$	$\frac{5}{6} \div 3 = ?$
$\frac{1}{3} \times \frac{1}{2} = \frac{1}{6}$	$\frac{1}{3} \div 2 = ?$
$\frac{4}{5} \times \frac{1}{7} = \frac{4}{35}$	$\frac{4}{5} \div 7 = ?$

$\frac{\overset{8}{\cancel{24}}}{\underset{1}{\cancel{5}}} \times \frac{\overset{2}{\cancel{10}}}{\underset{3}{\cancel{9}}} = \frac{16}{3} = 5\frac{1}{3}$	$\frac{24}{5} \div \frac{9}{10} = ?$
$\frac{\overset{2}{\cancel{14}}}{9} \times \frac{5}{\underset{1}{\cancel{7}}} = \frac{10}{9} = 1\frac{1}{9}$	$\frac{14}{9} \div \frac{7}{5} = ?$
$\frac{\overset{7}{\cancel{49}}}{5} \times \frac{2}{\underset{1}{\cancel{7}}} = \frac{14}{5} = 2\frac{4}{5}$	$\frac{49}{5} \div \frac{7}{2} = ?$
$\frac{\overset{1}{\cancel{3}}}{4} \times \frac{13}{\underset{2}{\cancel{6}}} = \frac{13}{8} = 1\frac{5}{8}$	$\frac{3}{4} \div \frac{6}{13} = ?$
$\frac{17}{\underset{3}{\cancel{12}}} \times \frac{\overset{1}{\cancel{4}}}{5} = \frac{17}{15} = 1\frac{2}{15}$	$\frac{17}{12} \div \frac{5}{4} = ?$

$\frac{4}{5}$ $\frac{6}{11}$	$2 \times \frac{2}{5} =$	$3 \times \frac{2}{11} =$
$\frac{1}{12}$ $\frac{5}{32}$	$\frac{1}{2} \div 6 = ?$	$\frac{5}{16} \div 2 = ?$
$\frac{5}{7}$ $\frac{2}{9}$	$5 \times \frac{1}{7} =$	$2 \times \frac{1}{9} =$
$\frac{7}{24}$ $\frac{9}{100}$	$\frac{7}{8} \div 3 = ?$	$\frac{9}{10} \div 10 = ?$
$\frac{12}{17}$ $\frac{18}{19}$	$3 \times \frac{4}{17} =$	$6 \times \frac{3}{19} =$

$\frac{\overset{4}{\cancel{16}}}{3} \times \frac{1}{\underset{1}{\cancel{4}}} = \frac{4}{3} = 1\frac{1}{3}$	$\frac{16}{3} \div 4 = ?$
$\frac{\overset{3}{\cancel{9}}}{\underset{2}{\cancel{8}}} \times \frac{\overset{1}{\cancel{4}}}{\underset{1}{\cancel{3}}} = \frac{3}{2} = 1\frac{1}{2}$	$\frac{9}{8} \div \frac{3}{4} = ?$
$\frac{\overset{3}{\cancel{12}}}{7} \times \frac{5}{\underset{1}{\cancel{4}}} = \frac{15}{7} = 2\frac{1}{7}$	$\frac{12}{7} \div \frac{4}{5} = ?$
$\frac{7}{\underset{1}{\cancel{4}}} \times \frac{\overset{2}{\cancel{8}}}{3} = \frac{14}{3} = 4\frac{2}{3}$	$\frac{7}{4} \div \frac{3}{8} = ?$
$\frac{\overset{6}{\cancel{18}}}{\underset{1}{\cancel{7}}} \times \frac{\overset{2}{\cancel{14}}}{\underset{5}{\cancel{15}}} = \frac{12}{5} = 2\frac{2}{5}$	$\frac{18}{7} \div \frac{15}{14} = ?$

Reducing Fractions

The top and bottom parts of a fraction are called its terms.

The numerator is the top term. → $\frac{1}{14}$ ← The denominator is the bottom term.

3

The terms of the fraction $\frac{3}{4}$ are the numerator ____

4

and the denominator ____.

11

The terms of the fraction $\frac{11}{12}$ are the numerator ____

12

and the denominator____.

$\frac{2}{4}$

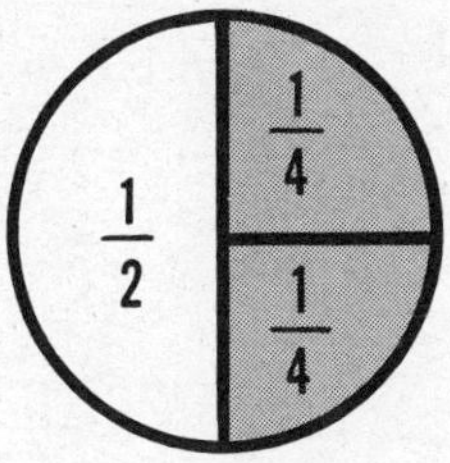

$\frac{1}{4} + \frac{1}{4} = \frac{2}{\_}$

$\frac{2}{4}$ of the circle is the same as $\frac{1}{2}$ of it.

so $\frac{2}{4} = \frac{1}{2}$

We have reduced the fraction $\frac{2}{4}$ to $\frac{1}{2}$.

To reduce a fraction, we divide both terms by the same number. This number must divide the numerator and the denominator evenly (without leaving a remainder).

We can divide 2 and 4 evenly by 2.

$\frac{1}{2}$

so $\frac{2}{4} = \frac{1}{\_}$ (4 ÷ 2)

$1\frac{1}{2}$	$\frac{3}{7} \div \frac{2}{7} = ?$ $\frac{3}{\underset{1}{\cancel{7}}} \times \frac{\overset{1}{\cancel{7}}}{2} = \frac{3}{2} =$
$\frac{8}{5} = 1\frac{3}{5}$	$\frac{6}{5} \div \frac{3}{4} = ?$ $\frac{\overset{2}{\cancel{6}}}{5} \times \frac{4}{\underset{1}{\cancel{3}}} = \frac{\quad}{\quad} =$
$\frac{7}{\underset{1}{\cancel{3}}} \times \frac{\overset{3}{\cancel{9}}}{2} = \frac{21}{2} = 10\frac{1}{2}$	$\frac{7}{3} \div \frac{2}{9} = ?$ $\frac{7}{3} \times \frac{9}{2} =$
$\frac{7}{4} = 1\frac{3}{4}$	$\frac{21}{16} \div \frac{3}{4} = ?$ $\frac{\overset{7}{\cancel{21}}}{\underset{4}{\cancel{16}}} \times \frac{\overset{1}{\cancel{4}}}{\underset{1}{\cancel{3}}} =$
$\frac{\cancel{7}}{\underset{4}{\cancel{8}}} \times \frac{\overset{5}{\cancel{10}}}{\cancel{7}} = \frac{5}{4} = 1\frac{1}{4}$	$\frac{7}{8} \div \frac{7}{10} = ?$

no

Let's try to reduce the fraction $\frac{2}{5}$. We can divide 2 evenly by 2. But $5 \div 2 = 2 \text{ r } 1$

Can we divide 5 evenly by 2? yes / no

We can divide 5 evenly only by 5.

---

no

Can we divide 2 evenly by 5? yes / no

The fraction $\frac{2}{5}$ cannot be reduced. We say that it is in lowest terms.

$\frac{4}{5}$

The fraction $\frac{4}{5}$ / $\frac{2}{4}$ is in lowest terms.

In each of the following pairs, circle the fraction that is in lowest terms:

Answer		
	(circled) $\frac{1}{3}$	$\frac{2}{6}$
$\frac{1}{4}$	$\frac{1}{4}$	$\frac{2}{8}$
$\frac{2}{3}$	$\frac{4}{6}$	$\frac{2}{3}$
$\frac{3}{5}$	$\frac{3}{5}$	$\frac{6}{10}$
$\frac{1}{6}$	$\frac{2}{12}$	$\frac{1}{6}$

$1\frac{2}{3}$	$\frac{1}{\underset{1}{\cancel{2}}} \times \frac{\overset{5}{\cancel{10}}}{3} = \frac{5}{3} =$
$\frac{7}{6} = 1\frac{1}{6}$	$\frac{\overset{1}{\cancel{2}}}{3} \times \frac{7}{\underset{2}{\cancel{4}}} = \text{—} =$
$\frac{3}{\underset{1}{\cancel{4}}} \times \frac{\overset{2}{\cancel{8}}}{5} = \frac{6}{5} = 1\frac{1}{5}$	$\frac{3}{4} \times \frac{8}{5} =$
$\frac{\overset{4}{\cancel{12}}}{7} \times \frac{2}{\underset{1}{\cancel{3}}} = \frac{8}{7} = 1\frac{1}{7}$	$\frac{12}{7} \times \frac{2}{3} =$

$1\frac{1}{2}$	Notice that we can sometimes reduce all 4 terms of a multiplication problem. $\frac{\overset{3}{\cancel{9}}}{\underset{2}{\cancel{16}}} \times \frac{\overset{1}{\cancel{8}}}{\underset{1}{\cancel{3}}} = \frac{3}{2} =$
$4\frac{1}{2}$	$\frac{\overset{3}{\cancel{15}}}{\underset{2}{\cancel{8}}} \times \frac{\overset{3}{\cancel{12}}}{\underset{1}{\cancel{5}}} = \frac{9}{2} =$
$\frac{\overset{7}{\cancel{49}}}{\underset{3}{\cancel{6}}} \times \frac{\overset{1}{\cancel{2}}}{\underset{1}{\cancel{7}}} = \frac{7}{3} = 2\frac{1}{3}$	$\frac{49}{6} \times \frac{2}{7} =$
$\frac{\overset{8}{\cancel{24}}}{\underset{1}{\cancel{5}}} \times \frac{\overset{4}{\cancel{20}}}{\underset{7}{\cancel{21}}} = \frac{32}{7} = 4\frac{4}{7}$	$\frac{24}{5} \times \frac{20}{21} =$

$\frac{2}{5}$	TEST
$\frac{6}{7}$	$2 \times \frac{1}{5} =$ $\frac{2}{7} \times 3 =$
$\frac{16}{19}$	$8 \times \frac{2}{19} =$ $\frac{4}{5} \times 5 =$
$\left(\frac{20}{5}\right) = 4$	
$\frac{3}{20}$	$\frac{3}{4} \div 5 = ?$ $\frac{5}{8} \div 2 = ?$
$\frac{5}{16}$	$\frac{3}{4} \times \frac{1}{5} =$
$\frac{7}{27}$	$\frac{7}{9} \div 3 = ?$ $\frac{3}{4} \div \frac{1}{4} = ?$
$\left(\frac{12}{4}\right) = 3$	
$\frac{1}{7}$	In each pair, circle the fraction that is in lowest terms: $\frac{1}{7}$ $\frac{2}{14}$
$\frac{7}{9}$	$\frac{14}{18}$ $\frac{7}{9}$

$1\frac{1}{5}$	Remember to reduce improper fractions to lowest terms and change them to mixed numbers. $\frac{19}{15} - \frac{1}{15} = \frac{18}{15} = \frac{6}{5} = 1\frac{1}{\quad}$
$1\frac{1}{2}$	$\frac{9}{4} - \frac{3}{4} = \frac{6}{4} = \frac{3}{2} =$
$\frac{5}{4} = 1\frac{1}{4}$	$\frac{23}{12} - \frac{8}{12} = \frac{15}{12} = \frac{\quad}{\quad} =$
$\frac{10}{6} = \frac{5}{3} = 1\frac{2}{3}$	$\frac{11}{6} - \frac{1}{6} = \frac{\quad}{6} = \frac{\quad}{\quad} =$
$\frac{34}{16} = \frac{17}{8} = 2\frac{1}{8}$	$\frac{43}{16} - \frac{9}{16} =$
$\frac{12}{8} = \frac{3}{2} = 1\frac{1}{2}$	$\frac{19}{8} - \frac{7}{8} =$
$\frac{24}{9} = \frac{8}{3} = 2\frac{2}{3}$	$\frac{31}{9} - \frac{7}{9} =$
$\frac{18}{14} = \frac{9}{7} = 1\frac{2}{7}$	$\frac{27}{14} - \frac{9}{14} =$
$\frac{22}{18} = \frac{11}{9} = 1\frac{2}{9}$	$\frac{35}{18} - \frac{13}{18} =$
$\frac{52}{24} = \frac{13}{6} = 2\frac{1}{6}$	$\frac{59}{24} - \frac{7}{24} =$

To reduce a fraction, we must always divide both terms <u>evenly</u> by the same number.

Reduce to lowest terms by dividing both terms by 2:

$\frac{2}{3}$	$\frac{4}{6} = \frac{2}{\_\_}$
$\frac{3}{4}$	$\frac{6}{8} = \frac{\_\_}{4}$
$\frac{4}{5}$	$\frac{8}{10} = \_\_$
$\frac{5}{6}$	$\frac{10}{12} = \_\_$
$\frac{6}{7}$	$\frac{12}{14} = \_\_$

Reduce to lowest terms by dividing both terms by 3:

$\frac{1}{2}$	$\frac{3}{6} = \frac{\_\_}{2}$
$\frac{1}{3}$	$\frac{3}{9} = \frac{1}{\_\_}$
$\frac{3}{4}$	$\frac{9}{12} = \_\_$
$\frac{1}{5}$	$\frac{3}{15} = \_\_$
$\frac{4}{7}$	$\frac{12}{21} = \_\_$

TEST

$\frac{3}{7} + \frac{5}{7} =$

$\frac{17}{5} - \frac{8}{5} =$

$\frac{7}{4} \times \frac{3}{2} =$

$\frac{1}{3} \div \frac{1}{8} =$?

Reduce to lowest terms and change to mixed numbers:

$\frac{30}{27} =$ $\frac{36}{30} =$

$\frac{90}{40} =$ $\frac{42}{12} =$

$\frac{4}{6} + \frac{5}{6} =$

$\frac{5}{8} + \frac{7}{8} + \frac{6}{8} =$

$\frac{8}{7} = 1\frac{1}{7}$

$\frac{9}{5} = 1\frac{4}{5}$

$\frac{21}{8} = 2\frac{5}{8}$

$\frac{1}{3} \times \frac{8}{1} = \frac{8}{3} = 2\frac{\;}{3}$

$\frac{10}{9} = 1\frac{1}{9}$

$\frac{6}{5} = 1\frac{1}{5}$

$\frac{9}{4} = 2\frac{1}{4}$

$\frac{7}{2} = 3\frac{1}{2}$

$\frac{9}{6} = \frac{3}{2} = 1\frac{1}{2}$

$\frac{18}{8} = \frac{9}{4} = 2\frac{1}{4}$

3	To reduce a proper fraction to lowest terms, first try dividing both terms by the numerator. The numerator of the fraction $\frac{3}{21}$ is ____.
$\frac{1}{7}$	$\frac{3}{21} = \frac{1}{}$ $\frac{(3 \div 3)}{(21 \div 3)}$
yes	Is $\frac{1}{7}$ in lowest terms? yes / no
no	$\frac{6}{15} = ?$ $15 \div 6 = 2 \text{ r } 3$ So we have to try the next smallest number that will divide the numerator 6 evenly. Is it 5? ($6 \div 5 = 1 \text{ r } 1$) yes / no
no	Is it 4? ($6 \div 4 = 1 \text{ r } 2$) yes / no
yes	Is it 3? ($6 \div 3 = 2$) yes / no
$\frac{2}{5}$	$\frac{6}{15} = \frac{2}{}$ ($15 \div 3$)
yes	Is $\frac{2}{5}$ in lowest terms? yes / no

Answers	Exercises
	Reduce each answer to lowest terms and then change it to a mixed number:
$1\frac{1}{2}$	$\frac{3}{4} + \frac{3}{4} = \frac{6}{4} = \frac{3}{2} = 1\frac{}{2}$
$1\frac{2}{3}$	$\frac{5}{6} + \frac{5}{6} = \frac{10}{6} = \frac{5}{3} =$
$\frac{5}{4} = 1\frac{1}{4}$	$\frac{7}{8} + \frac{3}{8} = \frac{10}{8} = \text{—} =$
$\frac{22}{14} = \frac{11}{7} = 1\frac{4}{7}$	$\frac{13}{14} + \frac{9}{14} = \text{—} = \text{—} =$
$\frac{16}{10} = \frac{8}{5} = 1\frac{3}{5}$	$\frac{9}{10} + \frac{7}{10} =$
$\frac{18}{12} = \frac{3}{2} = 1\frac{1}{2}$	$\frac{11}{12} + \frac{7}{12} =$
$\frac{28}{16} = \frac{7}{4} = 1\frac{3}{4}$	$\frac{13}{16} + \frac{15}{16} =$
$\frac{20}{14} = \frac{10}{7} = 1\frac{3}{7}$	$\frac{11}{14} + \frac{9}{14} =$
$\frac{21}{9} = \frac{7}{3} = 2\frac{1}{3}$	$\frac{8}{9} + \frac{8}{9} + \frac{5}{9} =$
$\frac{39}{18} = \frac{13}{6} = 2\frac{1}{6}$	$\frac{15}{18} + \frac{13}{18} + \frac{11}{18} =$

$\frac{3}{4}$	Reduce to lowest terms: $\frac{12}{16} = \frac{3}{\ }\ \frac{(12 \div 4)}{(16 \div 4)}$
$\frac{2}{3}$	$\frac{8}{12} = \frac{\ }{\ }$
$\frac{3}{5}$	$\frac{15}{25} = \frac{3}{\ }\ \frac{(15 \div 5)}{(25 \div 5)}$
$\frac{5}{8}$	$\frac{25}{40} = \frac{\ }{\ }$
$\frac{2}{3}$	$\frac{14}{21} = \frac{2}{\ }\ \frac{(14 \div 7)}{(21 \div 7)}$
$\frac{1}{3}$	$\frac{6}{18} = \frac{\ }{\ }$
$\frac{1}{2}$	$\frac{8}{16} = \frac{1}{\ }$
$\frac{1}{3}$	$\frac{9}{27} =$
$\frac{7}{8}$	$\frac{14}{16} =$
$\frac{1}{5}$	$\frac{4}{20} =$
$\frac{9}{10}$	$\frac{45}{50} =$

$\frac{5}{2} = 2\frac{1}{2}$	$\frac{10}{4} =$
$\frac{4}{3} = 1\frac{1}{3}$	$\frac{12}{9} =$
$\frac{8}{5} = 1\frac{3}{5}$	$\frac{16}{10} =$
$\frac{7}{6} = 1\frac{1}{6}$	$\frac{14}{12} =$
$\frac{5}{3} = 1\frac{2}{3}$	$\frac{15}{9} =$
$\frac{7}{5} = 1\frac{2}{5}$	$\frac{14}{10} =$
$\frac{8}{7} = 1\frac{1}{7}$	$\frac{16}{14} =$
$\frac{7}{3} = 2\frac{1}{3}$	$\frac{21}{9} =$
$\frac{6}{5} = 1\frac{1}{5}$	$\frac{12}{10} =$
$\frac{7}{6} = 1\frac{1}{6}$	$\frac{28}{24} =$
$\frac{9}{7} = 1\frac{2}{7}$	$\frac{45}{35} =$
$\frac{8}{3} = 2\frac{2}{3}$	$\frac{24}{9} =$

$\frac{1}{7}$

$\frac{3}{5}$

$\frac{1}{4}$

$\frac{1}{5}$

$\frac{5}{7}$

$\frac{1}{6}$

$\frac{3}{5}$

$\frac{5}{7}$

$\frac{4}{5}$

$\frac{7}{8}$

Reduce to lowest terms:

$\frac{2}{14} = \_\_\_$ $\qquad \frac{9}{15} = \_\_\_$

$\frac{4}{16} = \_\_\_$ $\qquad \frac{5}{25} = \_\_\_$

$\frac{15}{21} = \_\_\_$ $\qquad \frac{7}{42} = \_\_\_$

$\frac{24}{40} = \_\_\_$ $\qquad \frac{20}{28} = \_\_\_$

$\frac{36}{45} = \_\_\_$ $\qquad \frac{35}{40} = \_\_\_$

Answers	Problems	
	Circle the improper fraction that is in lowest terms:	
	$\frac{3}{2}$ (circled)	$\frac{6}{4}$
$\frac{4}{3}$	$\frac{8}{6}$	$\frac{4}{3}$
$\frac{8}{7}$	$\frac{16}{14}$	$\frac{8}{7}$
$\frac{3}{2}$	$\frac{3}{2}$	$\frac{9}{6}$
$\frac{6}{5}$	$\frac{6}{5}$	$\frac{18}{15}$
$\frac{9}{8}$	$\frac{27}{24}$	$\frac{9}{8}$

Reduce to lowest terms; then change to mixed numbers:

Answers	Problems
$1\frac{1}{3}$	
$1\frac{1}{2}$	$\frac{8}{6} = \frac{4}{3} = 1\frac{}{}$ —
$\frac{5}{4} = 1\frac{1}{4}$	$\frac{12}{8} = \frac{3}{2} =$
$\frac{9}{7} = 1\frac{2}{7}$	$\frac{25}{20} =$
$\frac{4}{3} = 1\frac{1}{3}$	$\frac{27}{21} =$
	$\frac{16}{12} =$

Always reduce answers to lowest terms.

$\frac{2}{6}$	$\frac{1}{6} + \frac{1}{6} = \frac{\quad}{6} = \frac{1}{3}$
$\frac{3}{4}$	$\frac{5}{8} + \frac{1}{8} = \frac{6}{8} = \frac{3}{\quad}$
$\frac{5}{10} = \frac{1}{2}$	$\frac{3}{10} + \frac{2}{10} = \frac{\quad}{10} = \frac{\quad}{2}$
$\frac{8}{12} = \frac{2}{3}$	$\frac{1}{12} + \frac{7}{12} = \frac{\quad}{\quad} = \frac{\quad}{\quad}$
$\frac{7}{14} = \frac{1}{2}$	$\frac{3}{14} + \frac{4}{14} = \frac{\quad}{\quad} = \frac{\quad}{\quad}$
$\frac{6}{9}$	$\frac{8}{9} - \frac{2}{9} = \frac{\quad}{9} = \frac{2}{3}$
$\frac{1}{2}$	$\frac{3}{4} - \frac{1}{4} = \frac{2}{4} = \frac{1}{\quad}$
$\frac{6}{16} = \frac{3}{8}$	$\frac{9}{16} - \frac{3}{16} = \frac{\quad}{16} = \frac{\quad}{8}$
$\frac{8}{20} = \frac{2}{5}$	$\frac{15}{20} - \frac{7}{20} = \frac{\quad}{\quad} = \frac{\quad}{\quad}$
$\frac{10}{18} = \frac{5}{9}$	$\frac{17}{18} - \frac{7}{18} = \frac{\quad}{\quad} = \frac{\quad}{\quad}$
$\frac{8}{28} = \frac{2}{7}$	$\frac{15}{28} - \frac{7}{28} = \frac{\quad}{\quad} = \frac{\quad}{\quad}$

$1\frac{2}{3}$	$\frac{10}{3} - \frac{5}{3} = \frac{5}{3} = 1\frac{\ }{}$ —
$\frac{11}{5} = 2\frac{1}{5}$	$\frac{17}{5} - \frac{6}{5} =$ —— $=$
$\frac{12}{7} = 1\frac{5}{7}$	$\frac{20}{7} - \frac{8}{7} =$
$\frac{22}{9} = 2\frac{4}{9}$	$\frac{29}{9} - \frac{7}{9} =$
$1\frac{1}{9}$	$\frac{5}{3} \times \frac{2}{3} = \frac{10}{9} =$
$\frac{25}{6} = 4\frac{1}{6}$	$\frac{5}{2} \times \frac{5}{3} =$
$\frac{21}{8} = 2\frac{5}{8}$	$\frac{7}{4} \times \frac{3}{2} =$
$3\frac{1}{3}$	$\frac{2}{3} \div \frac{1}{5} = ?$ $\frac{2}{3} \times \frac{5}{1} = \frac{10}{3} =$ ——
$\frac{3}{5} \times \frac{7}{2} = \frac{21}{10} = 2\frac{1}{10}$	$\frac{3}{5} \div \frac{2}{7} = ?$

TEST

$\frac{2}{3}$	$\frac{4}{9} + \frac{2}{9} =$
$\frac{1}{3}$	$\frac{5}{6} - \frac{3}{6} =$
$\frac{3}{4}$	$\frac{6}{12} + \frac{3}{12} =$
$\frac{1}{4}$	$\frac{7}{8} - \frac{5}{8} =$
$\frac{3}{8}$	$\frac{5}{16} + \frac{1}{16} =$
$\frac{3}{5}$	$\frac{9}{10} - \frac{3}{10} =$
$\frac{4}{5}$	$\frac{9}{15} + \frac{3}{15} =$
$\frac{2}{3}$	$\frac{13}{15} - \frac{3}{15} =$
$\frac{5}{6}$	$\frac{7}{18} + \frac{8}{18} =$
$\frac{1}{6}$	$\frac{7}{24} - \frac{3}{24} =$

	Any answer that is an improper fraction should be changed to a mixed number.
	$\frac{2}{3} + \frac{2}{3} = \frac{4}{3}$
$1\frac{1}{3}$	$\frac{4}{3} = 1\frac{1}{}$
$1\frac{2}{7}$	$\frac{5}{7} + \frac{4}{7} = \frac{9}{7} = 1\frac{}{7}$
$1\frac{2}{9}$	$\frac{4}{9} + \frac{7}{9} = \frac{11}{9} = 1$ —
$2\frac{1}{3}$	$\frac{5}{3} + \frac{2}{3} = \frac{7}{3} =$
$\frac{16}{5} = 3\frac{1}{5}$	$\frac{7}{5} + \frac{9}{5} = \frac{}{5} =$
$\frac{13}{3} = 4\frac{1}{3}$	$\frac{7}{3} + \frac{6}{3} =$
$\frac{15}{7} = 2\frac{1}{7}$	$\frac{9}{7} + \frac{6}{7} =$
$\frac{5}{4} = 1\frac{1}{4}$	$\frac{3}{4} + \frac{2}{4} =$
$\frac{19}{16} = 1\frac{3}{16}$	$\frac{11}{16} + \frac{8}{16} =$

$\frac{12}{78}$

$$\frac{12}{13} \times \frac{1}{6} = \frac{\quad}{78}$$

To reduce $\frac{12}{78}$, we divide both terms by 6:

$$\frac{12}{78} = \frac{\quad}{13}$$

$\frac{2}{13}$

We can simplify the problem by reducing some of the terms <u>before</u> we multiply.

We need to divide a numerator and a denominator by the same number.

$\frac{2}{13}$

We can divide this numerator by 6. → $\frac{12}{13} \times \frac{1}{6}$ ← We can also divide this denominator by 6.

$$(12 \div 6)\ \frac{\overset{2}{\cancel{12}}}{13} \times \frac{1}{\underset{1}{\cancel{6}}}\ (6 \div 6)$$

$$\text{so}\quad \frac{2}{13} \times \frac{1}{1} = \frac{\quad}{\quad}$$

$\frac{4}{7}$

$$(3 \div 3)\ \frac{2}{\underset{1}{\cancel{3}}} \times \frac{\overset{2}{\cancel{6}}}{7}\ (6 \div 3)$$

$$\text{so}\quad \frac{2}{1} \times \frac{2}{7} = \frac{\quad}{7}$$

TEST

Answers		
$16\frac{5}{6}$	$7\frac{1}{6} + 9\frac{4}{6}$	$12\frac{7}{15} + 84\frac{4}{15}$
$96\frac{11}{15}$		
$9\frac{1}{3}$	$17\frac{2}{3} - 8\frac{1}{3}$	$427\frac{5}{6} - 206\frac{4}{6}$
$221\frac{1}{6}$		

Change to mixed numbers:

Answers		
$1\frac{1}{5}$	$\frac{6}{5} = 1\frac{1}{\ }$	$\frac{21}{4} = 5\frac{\ }{4}$
$5\frac{1}{4}$		
$4\frac{2}{3}$	$\frac{14}{3} = \_\frac{2}{3}$	$\frac{7}{5} =$
$1\frac{2}{5}$		
$1\frac{5}{6}$	$\frac{11}{6} =$	$\frac{16}{3} =$
$5\frac{1}{3}$		
$5\frac{3}{4}$	$\frac{23}{4} =$	$\frac{8}{5} =$
$1\frac{3}{5}$		

	$\frac{\overset{1}{\cancel{4}}}{5} \times \frac{3}{\underset{2}{\cancel{8}}} = ?$
$\frac{3}{10}$	$\frac{1}{5} \times \frac{3}{2} = \frac{\quad}{10}$
$\frac{1}{18}$	$\frac{1}{\underset{6}{\cancel{12}}} \times \frac{\overset{1}{\cancel{2}}}{3} = \frac{\quad}{18}$
$\frac{3}{5}$	$\frac{\overset{1}{\cancel{4}}}{5} \times \frac{3}{\underset{1}{\cancel{4}}} = \text{——}$
$\frac{2}{35}$	$\frac{\overset{1}{\cancel{5}}}{7} \times \frac{2}{\underset{5}{\cancel{25}}} = \text{——}$
$\frac{4}{\underset{5}{\cancel{15}}} \times \frac{\overset{2}{\cancel{6}}}{5} = \frac{8}{25}$	$\frac{4}{15} \times \frac{6}{5} = \text{——}$
$\frac{\overset{2}{\cancel{6}}}{9} \times \frac{5}{\underset{3}{\cancel{9}}} = \frac{10}{27}$	$\frac{6}{9} \times \frac{5}{9} = \text{——}$
$\frac{3}{\underset{1}{\cancel{7}}} \times \frac{\overset{1}{\cancel{7}}}{8} = \frac{3}{8}$	$\frac{3}{7} \times \frac{7}{8} = \text{——}$

$1\frac{1}{3}$ $1\frac{2}{3}$	$\frac{4}{3} = 1\frac{\quad}{3}$	$\frac{5}{3} = 1\frac{2}{\quad}$
$2\frac{2}{3}$ $1\frac{1}{4}$	$\frac{8}{3} = \_\_\frac{2}{3}$	$\frac{5}{4} = 1\frac{1}{\quad}$
$2\frac{1}{3}$ $2\frac{1}{4}$	$\frac{7}{3} = \_\_\frac{\quad}{3}$	$\frac{9}{4} =$
$3\frac{2}{3}$ $3\frac{3}{4}$	$\frac{11}{3} =$	$\frac{15}{4} =$
$4\frac{1}{4}$ $3\frac{1}{3}$	$\frac{17}{4} =$	$\frac{10}{3} =$
$1\frac{3}{4}$ $2\frac{3}{4}$	$\frac{7}{4} =$	$\frac{11}{4} =$
$4\frac{1}{3}$ $4\frac{3}{4}$	$\frac{13}{3} =$	$\frac{19}{4} =$

$\frac{7}{15}$, $\frac{1}{16}$	$\frac{7}{12} \times \frac{4}{5} =$	$\frac{5}{8} \times \frac{1}{10} =$
$\frac{11}{20}$, $\frac{5}{39}$	$\frac{11}{18} \times \frac{9}{10} =$	$\frac{12}{13} \times \frac{5}{36} =$
$\frac{3}{8}$, $\frac{6}{35}$	$\frac{9}{21} \times \frac{7}{8} =$	$\frac{4}{7} \times \frac{3}{10} =$
$\frac{2}{7}$, $\frac{1}{15}$	$\frac{1}{3} \times \frac{6}{7} =$	$\frac{4}{5} \times \frac{1}{12} =$
$\frac{6}{49}$, $\frac{11}{24}$	$\frac{5}{7} \times \frac{6}{35} =$	$\frac{11}{21} \times \frac{7}{8} =$
$\frac{14}{31}$, $\frac{21}{29}$	$\frac{2}{3} \times \frac{21}{31} =$	$\frac{7}{8} \times \frac{24}{29} =$

Answers	Frames
4	We have changed improper fractions to whole numbers: $\frac{6}{2} = 3$ $\frac{12}{3} =$
3 r 1	Some improper fractions can be changed to mixed numbers: $\frac{7}{2} = ?$ $7 \div 2 = \_\_ r\,1$ We write the remainder 1 over the denominator: $\frac{7}{2} = 3\frac{1}{2} \leftarrow$
$1\frac{1}{2}$	$\frac{5}{2} = 2\frac{1}{2}$ $\frac{3}{2} = 1\frac{}{2}$
$4\frac{1}{2}$ $6\frac{1}{2}$	$\frac{9}{2} = \_\_\frac{1}{2}$ $\frac{13}{2} = 6\frac{1}{}$
$5\frac{1}{2}$ $8\frac{1}{2}$	$\frac{11}{2} = 5\,\_\_$ $\frac{17}{2} = 8\,\_\_$
$7\frac{1}{2}$ $9\frac{1}{2}$	$\frac{15}{2} = \_\_\frac{1}{2}$ $\frac{19}{2} =$

$\frac{3}{1}$	$\frac{1}{6} \div \frac{1}{3} = ?$
$\frac{1}{2}$	$\frac{1}{3}$ inverted is —.
	$\frac{1}{\cancel{6}_{2}} \times \frac{\cancel{3}^{1}}{1} = \frac{1}{\_\_}$
$\frac{1}{4}$	$\frac{4}{7} \div 4 = ?$
$\frac{1}{7}$	4 inverted is ___.
	$\frac{\cancel{4}^{1}}{7} \times \frac{1}{\cancel{4}_{1}} = \frac{\_\_}{7}$
$\frac{\cancel{2}^{1}}{5} \times \frac{1}{\cancel{2}_{1}} = \frac{1}{5}$	$\frac{2}{5} \div 2 = ?$ $\frac{5}{16} \div \frac{3}{8} = ?$
$\frac{5}{\cancel{16}_{2}} \times \frac{\cancel{8}^{1}}{3} = \frac{5}{6}$	$\frac{2}{5} \times \frac{1}{2} = \_\_\_$ $\frac{5}{16} \times \frac{8}{3} = \_\_\_$
$\frac{7}{\cancel{12}_{3}} \times \frac{\cancel{4}^{1}}{3} = \frac{7}{9}$	$\frac{7}{12} \div \frac{3}{4} = ?$ $\frac{10}{11} \div 5 = ?$
$\frac{\cancel{10}^{2}}{11} \times \frac{1}{\cancel{5}_{1}} = \frac{2}{11}$	
$\frac{\cancel{2}^{1}}{3} \times \frac{5}{\cancel{4}_{2}} = \frac{5}{6}$	$\frac{2}{3} \div \frac{4}{5} = ?$ $\frac{7}{8} \div 14 = ?$
$\frac{\cancel{7}^{1}}{8} \times \frac{1}{\cancel{14}_{2}} = \frac{1}{16}$	

$\frac{3}{7}$	$6\frac{5}{7} - 4\frac{2}{7} = ?$ Subtract the fractions first. $\frac{5}{7} - \frac{2}{7} = \frac{}{7}$
$2\frac{3}{7}$	$6\frac{5}{7} - 4\frac{2}{7} = \frac{3}{7}$
$4\frac{3}{5}$	$7\frac{4}{5} - 3\frac{1}{5} = \frac{3}{5}$ $9\frac{7}{9} - 2\frac{5}{9}$
$7\frac{2}{9}$	
$3\frac{1}{8}$	$14\frac{7}{8} - 11\frac{6}{8}$ $23\frac{8}{10} - 17\frac{5}{10}$
$6\frac{3}{10}$	
$19\frac{5}{11}$	$38\frac{8}{11} - 19\frac{3}{11}$ $52\frac{12}{13} - 36\frac{8}{13}$
$16\frac{4}{13}$	
$25\frac{11}{15}$	$109\frac{13}{15} - 84\frac{2}{15}$ $375\frac{14}{17} - 189\frac{5}{17}$
$186\frac{9}{17}$	

$\frac{1}{2}$

$\frac{9}{15} = \frac{3}{5}$

$\frac{5}{25} = \frac{1}{5}$

$\frac{4}{24} = \frac{1}{6}$

$\frac{2}{7}$

$\frac{\overset{1}{\cancel{4}}}{5} \times \frac{1}{\underset{3}{\cancel{12}}} = \frac{1}{15}$

$\frac{\overset{1}{\cancel{5}}}{7} \times \frac{6}{\underset{7}{\cancel{35}}} = \frac{6}{49}$

$\frac{11}{\underset{3}{\cancel{21}}} \times \frac{\overset{1}{\cancel{7}}}{8} = \frac{11}{24}$

$\frac{\overset{2}{\cancel{20}}}{31} \times \frac{1}{\underset{1}{\cancel{10}}} = \frac{2}{31}$

$\frac{9}{\underset{2}{\cancel{24}}} \times \frac{\overset{1}{\cancel{12}}}{7} = \frac{9}{14}$

$\frac{5}{16} + \frac{3}{16} = \frac{8}{16} = \underline{\quad\quad}$

$\frac{2}{15} + \frac{7}{15} = \underline{\quad\quad} = \underline{\quad\quad}$

$\frac{8}{25} - \frac{3}{25} = \underline{\quad\quad} = \underline{\quad\quad}$

$\frac{11}{24} - \frac{7}{24} = \underline{\quad\quad} = \underline{\quad\quad}$

$\frac{1}{\cancel{3}} \times \frac{\cancel{6}}{7} = \underline{\quad\quad}$ $\quad\quad$ $\frac{4}{5} \times \frac{1}{12} =$

$\frac{5}{7} \times \frac{6}{35} = \underline{\quad\quad}$ $\quad\quad$ $\frac{11}{21} \times \frac{7}{8} =$

$\frac{20}{31} \div 10 = ?$ $\quad\quad$ $\frac{9}{24} \div \frac{7}{12} = ?$

Answers	Problem	Problem
$7\frac{2}{3}$ $9\frac{5}{7}$	$3\frac{1}{3} + 4\frac{1}{3} = \_\frac{2}{3}$	$5\frac{2}{7} + 4\frac{3}{7} = \_\frac{5}{}$
$8\frac{3}{4}$ $13\frac{5}{8}$	$3\frac{1}{4} + 5\frac{2}{4}$	$6\frac{3}{8} + 7\frac{2}{8}$
$40\frac{7}{9}$ $64\frac{9}{11}$	$17\frac{2}{9} + 23\frac{5}{9}$	$36\frac{3}{11} + 28\frac{6}{11}$
$20\frac{9}{10}$ $33\frac{7}{12}$	$8\frac{1}{10} + 5\frac{3}{10} + 7\frac{5}{10}$	$6\frac{5}{12} + 9\frac{1}{12} + 18\frac{1}{12}$
$92\frac{4}{5}$ $102\frac{12}{13}$	$37\frac{1}{5} + 42\frac{2}{5} + 13\frac{1}{5}$	$15\frac{3}{13} + 24\frac{4}{13} + 63\frac{5}{13}$

FINAL EXAM - Side 1

$\frac{4}{7} + \frac{2}{7} =$ $\qquad$ $\frac{3}{11} + \frac{5}{11} =$ $\qquad$ $\frac{6}{19} + \frac{7}{19} =$

$\frac{5}{9} - \frac{3}{9} =$ $\qquad$ $\frac{15}{21} - \frac{11}{21} =$ $\qquad$ $\frac{4}{5} - \frac{3}{5} =$

$\frac{3}{4} \times \frac{1}{5} =$ $\qquad$ $\frac{5}{6} \times \frac{3}{4} =$ $\qquad$ $\frac{7}{9} \times \frac{2}{3} =$

$\frac{5}{7} \div 3 = ?$ $\qquad$ $\frac{4}{5} \div \frac{7}{8} = ?$ $\qquad$ $\frac{6}{11} \div \frac{3}{4} = ?$

Reduce to lowest terms:

$\frac{12}{20} =$ $\qquad$ $\frac{10}{30} =$ $\qquad$ $\frac{21}{27} =$

$\frac{15}{16} \times \frac{3}{5} =$ $\qquad$ $\frac{6}{21} \times \frac{7}{8} =$

$\frac{8}{9} \div 4 = ?$ $\qquad$ $\frac{5}{7} \div \frac{11}{14} = ?$ $\qquad$ $\frac{9}{17} \div \frac{3}{5} = ?$

Turn the book over and continue with page 49.

Mixed Numbers

$2\frac{1}{3}$ is a mixed number. It is made up of

__ a. just a whole number.
__ b. just a fraction.
__ c. a whole number and a fraction.

✓ c.

Circle the mixed numbers.

$6\frac{7}{8}$ (circled) 9 $\frac{3}{4}$ $5\frac{1}{2}$ $\frac{14}{19}$ 12 $10\frac{8}{9}$

$5\frac{1}{2}$ $10\frac{8}{9}$

$$\begin{array}{r} 2\frac{1}{5} \\ +4\frac{1}{5} \\ \hline ? \end{array}$$

Add the fractions first: $\frac{1}{5} + \frac{1}{5} = \frac{\ }{5}$

$\frac{2}{5}$

$$\begin{array}{r} 2\frac{1}{5} \\ +4\frac{1}{5} \\ \hline ?\ \frac{2}{5} \end{array}$$

Now add the whole numbers: 2 + 4 = ___

6

$$\begin{array}{r} 2\frac{1}{5} \\ +4\frac{1}{5} \\ \hline 6\frac{2}{5} \end{array}$$